AF541547

Introductory Agricultural Meteorology

NIPA® GENX ELECTRONIC RESOURCES & SOLUTIONS P. LTD.
New Delhi-110 034

About the Authors

Surender Singh has about thirty five years of experience with PhD in the field of Agrometeorology. Worked in Agrometeorology encompassing teaching, research, extension and administration at various capacities *viz*., Principal Scientist, Professor & Head, Senior Scientist, Assistant Scientist, Advisor (Recruitments) and Associate Director (Students' Counseling and Placement) at CCS Haryana Agricultural University, Hisar, India. Recipient of Two Gold Medals for Best Research Work during PhD Program and for Best PhD Dissertation (2001-02) awarded by then President of India Dr APJ Abdul Kalam. Offered UK Universities Vice Chancellors' Scholarship for Higher Studies twice (1991 & 1993) under Commonwealth Fellowship Program. Post Doctoral Fellow at Abdus Salam International Centre for Theoretical Physics, Trieste, Italy; IAC – Instituto Agronômico, Campinas, Brazil and BOKU (Met) - Institute of Meteorology and Climatology, Vienna, Austria.

He is life member of Association of Agrometeorologists, Indian Meteorological Society, Indian Society of Remote Sensing, International Society for Agrometeorology, Society for Sustainable Agric& Resource Management, Environment Science Published for Everybody Round the Earth (ESPERE), European Drought Centre, South Asian Forum on Agriculture Meteorology (SAFOAM) and The Geographical Society of India. Organized quite a few trainings/symposia in Agrometeorology and associated domains at National level. Published more than 250 research papers, popular articles, manuals/teaching aids and research bulletins/book chapters etc. Performed responsibilities as Managing Editor - Journal of Agrometeorology; Research Collaborator, Monsoon Asia Agro-Environmental Research Consortium (MARCO), Japan; SPARC Research Collaborator, ETH, Zurich, Switzerland; Member, Expert Team - CAgM-WMO, Geneva, Switzerland (2018-20). Core Committee Member on Agriculture – Drought at WMO, Geneva, Switzerland (2021-24 & 2024-27). External Expert for evaluation of research projects, COST-Association, European Union, Brussels, Belgium. Many students (UG, PG and PhD) have been trained under him and placed across the country and well settled in various positions in the field of Agrometeorology. He is widely travelled researcher/expert across the globe on many academic pursuits. He has been conferred upon Fellow of Association of Agrometeorologists (FAAM) for the year 2024 in recognition of his exemplary achievements and enduring impact on agrometeorological science.

V Uma Maheswara Rao has forty years of experience with PhD in the broader field of Agrometeorology. Worked in Agrometeorology encompassing teaching, research, extension and administration at various capacities viz., Project Coordinator, Principal Scientist, Professor & Head, Agrometeorologist, and Assistant Scientist at CCS Haryana Agricultural University and Indian Council of Agriculture Research in India.

He is life member of Indian Meteorological Society, The Arid Zone Research Association of India, Indian Society of Remote Sensing, Association of Agrometeorologists, International Society for Agrometeorology, and The Indian Society of Dryland Agriculture. Organized several trainings/workshops in Agrometeorology at the International and National level. Published more than 345 research papers, popular articles, manuals/teaching aids and research bulletins/book chapters. Member of the CCI-ET-IIC, WMO, Geneva (2014-2017). Member of XI Plan High Powered Review Committee of Ministry of Earth Sciences, Government of India (2012). In 2014, the prestigious Chaudhary Devi Lal Award, ICAR's Best AICRP on Agrometeorology was received under his dynamic leadership as Project Coordinator. Many students (UG, PG and PhD) have been trained under his guidance are placed across the country and settled in various positions in the field of Agrometeorology. He visited numerous countries in official capacity.

BV Ramana Rao taught Agricultural Meteorology in the University of Agricultural Sciences, Bangalore during the years 1966 to 1979 and later joined the Indian Council of Agricultural Research and superannuated from service during April, 2000. He served as Head, Division of Climatology, Wind Power and Solar Energy Utilization at ICAR-Central Arid Zone Research Institute, Jodhpur (Rajasthan); founder Project Coordinator, All India Coordinated Research Project on Agrometeorology at ICAR Central Research Institute for Dryland Agriculture, Hyderabad (Telangana) during December 1984 to March 1994. His major contributions in the field of Agrometeorology include Agroclimatic Characterization, Delineation of Efficient Cropping Zones, Crop Weather Modeling, Classification and Management of Agricultural Droughts, Operational Agrometeorology, etc. He published about 150 research papers in peer reviewed journals with 50 of his papers appearing in high impact International Journals. He authored 15 books and monographs and a number of scientific and technical bulletins. He served as Editor in Chief, Journal of Agrometeorology during the years 2009 to 2023.

He is the recipient of British Council Fellowship during the year 1986-87 for advanced training at University of Reading, United Kingdom. He was conferred as Honorary Fellow (2011) and Life Time Achievement Award (2025) by the Association of Agrometeorologists for his valuable contribution to the growth and development of Agricultural Meteorology in India.

Introductory Agricultural Meteorology

Surender Singh
MSc; PhD; FAAM
Professor/Principal Scientist (Agrometeorology)
CCS HAU, Hisar, Haryana, India

V Uma Maheswara Rao
MSc; PGDIAS; PhD; FAAM
Former Project Coordinator - AICRP on Agrometeorology
ICAR-CRIDA, Hyderabad, Telangana, India

BV Ramana Rao
MSc Tech; FAAM
Founder Project Coordinator - AICRP on Agrometeorology
ICAR-CRIDA, Hyderabad, Telangana, India

NIPA® GENX ELECTRONIC RESOURCES & SOLUTIONS P. LTD.
New Delhi-110 034

NIPA® GENX ELECTRONIC
RESOURCES & SOLUTIONS P. LTD.

101,103, Vikas Surya Plaza, CU Block
L.S.C. Market, Pitam Pura, New Delhi-110 034
Ph : +91-11-43860225, Mob.: +91 9717133558, 9540816132
E-mail: newindiapublishingagency@gmail.com
Website: www.nipaersources.com

Print ISBN: 978-93-58871-48-7
ebook ISBN: 978-93-58878-70-7

Composed and Designed by NIPA®.

The Book

Introductory Agricultural Meteorology

is dedicated to

Dr. PD Mistry

(1921- 2017)

Dr Purachand Dalpatram Mistry was among the early pioneers in India who recognized the need for a deeper integration of meteorology with agricultural practices. He played critical role in advancing meteorological education within the field of agriculture. He was deeply involved in curriculum development and the promotion of agricultural meteorology as an essential subject within agricultural universities. His advocacy led to the integration of meteorological principles into the academic training of agricultural scientists, horticulturists, and agronomists. This has had long-term benefits, equipping future generations of professionals with the tools needed to better manage agricultural production in the face of climate variability.

Dr Mistry's contributions to agricultural meteorology have left a lasting legacy in India and abroad. His work has empowered farmers with the tools and knowledge to make informed decisions based on weather data, and his research has played a pivotal role in enhancing the scientific understanding of how weather and climate influence agriculture. The importance of his work continues to grow, especially as the world faces the challenges posed by climate change and its impact on food security. His pioneering research, combined with his advocacy for the integration of meteorology and agriculture, has inspired many generations of scientists, students, and agricultural professionals to approach farming in a way that is both scientifically informed and environmentally sustainable.

Dr N Chattopadhyay

Adviser – Agrometeorology
The World Bank

Chairman (ET-GAMP)
World Meteorological Organization

Foreword

Agriculture is the primary national concernand is intricately linked with the natural environment.Every aspect of agriculture from long term planning to tactical decisions in day-to-day agricultural operations is dependent on climate and weather. Thus, in increasing food production and attaining self-sufficiency in agricultural production, meteorology has a very vital role to play. In order to keep pace with the increasing population, the growth in agricultural production should be sustainable in the long run.An increase in food production has become a matter of vital importance in keeping pace with the increasing population. Besides,uncertainties of weather and climate pose a major threat to food security.There can be hardly any doubt that there will be increasingly greater application of scientific methods of application of weather and climate to agriculture and farming in future.Imparting education with a focus on skill development and professional excellence is crucial for meeting the complex challenges facing the world today. The researchable issues are many in the field of climate change and crop production, while the achievements so far in this important area are much less. There are more such issues directly or indirectly related to the food production have to addressed with clear cut solution. Recognizing these challenges, the Indian Council of Agricultural Research (ICAR) has taken a significant step forward by developing a new syllabus based on the recommendations of the 6th Deans Committee. This syllabus is designed to equip students with the knowledge, skills, and competencies required to address the emerging issues in agriculture, horticulture, sericulture, and other allied sectors..

Knowledge of the influence of weather on crops is as old as the science of meteorology. While weather is the ultimate result of a number of interacting physical processes in ocean and atmosphere, crop growth and yield are the resultant of a number of physiological processes in the plant. In depth understanding of the crop weather relationship is indispensable to make a sustainable agricultural platform. Agricultural meteorology plays a vital role in this endeavor, providing critical insights into the complex relationships between weather, climate, and agricultural systems. By understanding these relationships, farmers, researchers, and policymakers can make informed decisions that enhance crop productivity, reduce risks, and promote sustainable agricultural practices. It is in this context that the book "Introductory Agricultural Meteorology" makes a valuable contribution.

The authors, Professors Surender Singh, V Uma Maheswara Rao and BV Ramana Rao, are well-known experts in the discipline of agricultural meteorology, with a deep understanding of the subject and its applications. Their book provides a detailed explanation of the basic principles of meteorology, weather forecasting, climate change, and the application of agricultural meteorology in agriculture, horticulture, and sericulture. The text is designed to provide students with a better learning experience, combining theoretical knowledge with practical examples and case studies.

It is highly gratifying to note that the authors have made commendable efforts in developing a textbook that meets the needs of students and faculty in agricultural meteorology. Their contribution will undoubtedly enhance the understanding and application of agricultural meteorology in India and beyond. By providing a comprehensive and accessible resource, the authors have opened up new avenues for learning and skill development, empowering students to become professionals who can drive innovation and sustainability in agriculture and allied sectors.

As we face a future marked by both challenges and possibilities, this text will serve as a compass, guiding future generations toward innovative solutions that ensure the continued prosperity of global agriculture.The purpose of this book is to offer perspectives on more productive, sustain able and resilient modes of agriculture.I wish the book all success and believe it will become a key resource for those embarking on careers in agricultural science, environmental studies, and related disciplines. It provides a foundation for understanding the impact of meteorological factors on farming, equipping readers with the knowledge to address challenges and seize opportunities in the evolving field of agricultural meteorology.

(Nabansu Chattopadhyay)
President, International Society for Agricultural Meteorology (INSAM)
Former Dy Director General Meteorology &
Head, Agricultural Meteorology Division, India Meteorological Department

Preface

Agriculture, being the backbone of economies and livelihoods across the world, is profoundly influenced by the forces of nature, particularly weather and climate. In India, where agriculture remains a central aspect of daily life and national development, understanding the interplay between atmospheric phenomena and agricultural practices is critical to ensuring both sustainable farming and food security.

Agricultural meteorology is a specialized field of study that bridges the gap between atmospheric sciences and practical agriculture. As global agricultural practices evolve in response to changing climates, the importance of understanding meteorological principles and their application to crop, horticulture, and sericulture production has never been more critical.

In this context, the Indian Council of Agricultural Research (ICAR), based on the recommendations of the 6th Deans Committee report, has outlined a course content that aims to equip students and faculty with the knowledge and skills necessary to harness the power of meteorology in agriculture, horticulture, and sericulture. This book, "Introductory Agricultural Meteorology," is designed to fulfill this objective, providing a thorough grounding in the principles of meteorology, weather forecasting, climate change, and the application of agricultural meteorology in these critical sectors.

The book is structured to cover the fundamental aspects of meteorology, including the Earth's atmosphere, weather systems, and climate dynamics. It delves into the intricacies of weather forecasting, exploring the various tools and techniques used to predict weather patterns and mitigate the impacts of adverse weather conditions. The text also examines the pressing issue of climate change, its causes, consequences, and implications for agricultural systems, as well as strategies for mitigating and adapting to these changes.

What sets this book apart is its focus on the practical applications of agricultural meteorology in agriculture, horticulture, and sericulture. Through a combination of theoretical explanations, case studies, and real-world examples, the book illustrates how meteorological knowledge can be used to optimize crop yields, improve water management, and reduce the risks associated with weather-related disasters.

This book has been developed with two primary objectives in mind: to provide faculty with a comprehensive resource that can inform their teaching and research, and to offer students an engaging and accessible learning experience. By covering the course content prescribed by ICAR, this book aims to bridge the gap between theoretical knowledge and practical application, empowering students and professionals to make informed decisions and drive innovation in agriculture, horticulture, and sericulture.

By exploring the complex relationships between weather, climate, and agricultural systems, this book seeks to inspire a new generation of professionals, researchers, and practitioners who can harness the power of meteorology to build more resilient, productive, and sustainable agricultural systems. It is our hope that this book will contribute to the development of a more climate-smart agriculture, one that prioritizes the well-being of farmers, ecosystems, and communities alike. We are grateful to Dr Nabansu Chattopadhyay, Chairman of Expert Team on Guide to Agricultural Meteorological Practices (ET-GAMP), World Meteorological Organisation and Former Deputy Director General Meteorology & Head, Agricultural Meteorology Division, India Meteorological Department, MoES, Govt of India for writing foreword with a message that education is no longer a process of acquiring knowledge but application of scientific knowledge to empower the farming community for adoption of latest technologies for prosperity.

It is our hope that this textbook will not only serve as a valuable academic resource but also inspire students to become leaders in the fields of agriculture, horticulture, and sericulture, contributing to the advancement of science and the sustainability of agriculture in an ever-changing climate. We welcome suggestions and comments from the readers of the book for improvements in future.

Surender Singh
V Uma Maheswara Rao
BV Ramana Rao

Acknowledgements

The creation of this textbook, *Introductory Agricultural Meteorology* has been an enriching and collaborative journey. It would not have been possible without the guidance, support, and contributions of many individuals and institutions.

First and foremost, we express our deepest gratitude to the Indian Council of Agricultural Research (ICAR) for their valuable guidance and for formulating the 6th Deans' Committee syllabus that has provided the framework for this book. Their vision for modernizing agricultural education and fostering an interdisciplinary approach has been instrumental in shaping the content and structure of this textbook.

We would like to extend our sincere appreciation to the faculty and experts from various disciplines of agriculture, horticulture, and sericulture, whose valuable feedback and insights have enriched the content. Their continuous support throughout the writing process has been essential in ensuring the accuracy and relevance of the material.

Special thanks to the students and practitioners in the field of agricultural meteorology whose questions, enthusiasm, and desire to understand the complexities of meteorological factors in farming have inspired much of the content. Their engagement in practical applications of meteorological knowledge reinforces the importance of bridging theory with real-world solutions.

We are also grateful to our colleagues and mentors who have provided constant encouragement, sharing their expertise and knowledge in meteorology, agriculture, and climate science. Their contributions, both technical and moral, have been a great source of motivation.

Heartfelt thanks goes to the editorial and publishing team for their tireless efforts in bringing this book to life. Their professionalism, attention to detail, and prompt responses to every query made this journey smooth and efficient.

We would also like to acknowledge the role of our family and friends for their unwavering support, patience, and understanding throughout the course of writing this book. Their belief in the importance of this work has been a constant source of strength.

Lastly, we dedicate this book to the students of agriculture, horticulture, and sericulture. It is our sincere hope that this textbook will not only enhance their understanding of agricultural meteorology but also inspire them to contribute meaningfully to the advancement of sustainable agricultural practices in an increasingly unpredictable world.

Surender Singh
V Uma Maheswara Rao
BV Ramana Rao

Contents

List of Figures

1

Introduction to Meteorology and Agricultural Meteorology

Meteorology is the scientific study of the **atmosphere**. It focuses on the physical processes and phenomena that occur within the Earth's atmosphere, including weather and climate. Meteorologists study these processes to understand, explain, and ultimately forecast the Earth's weather.

Think of the atmosphere as a vast and dynamic fluid that surrounds our planet (Fig. 1.1). Meteorology seeks to understand:

- **Composition of the atmosphere:** What gases and particles make up the air we breathe?
- **Structure of the atmosphere:** How is the atmosphere layered (e.g., troposphere, stratosphere)?
- **Atmospheric motion:** Why does the wind blow? What drives air currents?
- **Energy balance:** How does the Earth receive and redistribute energy from the sun?
- **Formation of weather phenomena:** How do clouds, rain, snow, storms, and other weather events develop?

Fig. 1.1: Atmosphere - A vast and dynamic fluid that surrounds planet earth

1.1 Importance of Meteorology in General

Beyond its crucial role in **agriculture**, understanding of meteorology is vital for:

- **Public safety:** Forecasting severe weather events like hurricanes, tornadoes, and heatwaves helps save lives and property.
- **Transportation:** Aviation, shipping, and land transportation rely heavily on weather forecasts for safe and efficient operations.
- **Water resource management:** Understanding precipitation patterns is essential for managing water supplies.
- **Energy production:** Wind and solar energy are directly influenced by atmospheric conditions.
- **Environmental studies:** Meteorology plays a role in understanding air pollution and climate change.

1.1.1. Factors Controlling Food Production: Food is Absolutely Fundamental to Life. It Provides us with:

- **Energy:** To carry out daily activities, work, and even think. Carbohydrates, fats, and proteins are our primary sources of energy.
- **Nutrients:** Essential substances like vitamins, minerals, proteins, fats, and carbohydrates that our bodies need to grow, repair tissues, and function properly.
- **Growth and Development:** Especially crucial for children and adolescents, food provides the building blocks for growth.
- **Protection against diseases:** A balanced diet strengthens our immune system, making us less susceptible to illnesses.
- **Mental Well-being:** Good nutrition is linked to better mood, concentration, and overall mental health.

In essence, food is not just about survival; it's about living a healthy and fulfilling life.

1.1.2. Factors Affecting Food Production

The production of food, primarily through agriculture, is a complex process influenced by a wide range of factors. These can be broadly categorized as:

- **Environmental/Physical Factors**
    - **Climate:** Temperature, rainfall, sunlight, humidity, and wind directly impact crop growth and livestock health.
    - **Soil:** Soil type, fertility, structure, and drainage are crucial for plant growth.

- **Water Availability:** Adequate water for irrigation and livestock is essential.
- **Topography:** The shape and features of the land can affect cultivation practices and suitability for different types of agriculture.
- **Pests and Diseases:** These can significantly reduce yields in both crops and livestock.

- **Biological Factors**
  - **Plant Varieties and Animal Breeds:** The genetic potential for yield, disease resistance, and other desirable traits varies greatly.
  - **Weeds:** Competition from weeds reduces the resources available for crops.
  - **Microorganisms:** Soil microbes play a role in nutrient cycling, while others can cause diseases.
- **Socio-economic Factors**
  - **Technology:** Availability and adoption of modern farming techniques, machinery, and biotechnology.
  - **Infrastructure:** Roads, storage facilities, and market access play a vital role in getting food from farms to consumers.
  - **Government Policies:** Subsidies, regulations, and trade agreements can significantly influence agricultural production.
  - **Economic Conditions:** Market prices, input costs (seeds, fertilizers, etc.), and access to credit affect farmers' decisions.
  - **Labor Availability:** The workforce available for agricultural activities.
  - **Land Ownership and Tenure:** How land is owned and managed impacts investment and practices.
- Factors We Have Control Over

While many factors influence food production, we have varying degrees of control over them:

- **Choice of crops and animal breeds:** Farmers can select varieties and breeds suited to their environment and market demands.
- **Farming practices:** Tillage methods, irrigation techniques, fertilization, pest and disease management, and crop rotation are largely within our control.
- **Technology adoption:** Farmers can choose to adopt new technologies to improve efficiency and yields.

    - **Soil management:** Practices like adding organic matter, conservation tillage, and managing soil pH can improve soil health.
    - **Water management:** Implementing efficient irrigation systems and water conservation practices.
    - **Post-harvest management:** Proper storage and handling can reduce food losses.
- **Factors We Do Not Have Direct Control Over**

Some factors are largely outside of our direct control at the individual farm level:

    - **Climate and weather:** While we can try to adapt, we cannot directly control rainfall, temperature, or extreme weather events.
    - **Soil type (inherent):** The basic geological origin of the soil is fixed. However, we can influence its properties.
    - **Global market prices:** These are influenced by worldwide supply and demand.
    - **Major natural disasters:** Floods, droughts, and other large-scale events can have devastating impacts.
    - **Some regional pests and diseases:** While we can manage them on our farms, their initial occurrence and spread can be influenced by broader environmental factors.
    - **Government policies (at an individual level):** Farmers must generally operate within the existing policy framework.

Understanding these controllable and uncontrollable factors is crucial for developing strategies to enhance food production and ensure food security.

1.2. Agriculture

Agriculture is the science, art, and practice of cultivating the soil, producing crops, and raising livestock for food, fiber, and other products. To deal with uncontrollable factors like **weather and climate** in agriculture, we focus on **adaptation** through:

- **Choosing crops based on climatic conditions:** Selecting plant varieties and animal breeds that are well-suited to the prevailing temperature, rainfall patterns, and other aspects of the local climate.
- **Understanding the effect of changing weather on crop growth, development, and yield:** Recognizing how variations in temperature, rainfall, sunlight, and other weather elements at different growth stages impact agricultural productivity.

- **Managing agricultural operations based on weather forecasts:** Utilizing short-term and long-term weather predictions to optimize planting, irrigation, fertilization, pest and disease management, and harvesting schedules.
- Employing **water-efficient irrigation** techniques.
- Practicing **conservation agriculture** to enhance soil moisture retention.
- Considering **crop insurance** to buffer against weather-related losses.
- Diversifying **crops and livestock** to reduce vulnerability.
- Utilizing **protected cultivation** (e.g., greenhouses, shade nets) through modified microclimate.

1.2.1. Managing agricultural operations based on weather forecasts

It involves strategically planning and executing farming activities by taking into account predicted weather conditions. This can significantly optimize resource use, reduce risks, and improve overall efficiency. Here are some examples:

- **Planting:** Delaying planting if heavy rainfall is forecast that could lead to waterlogging and seed damage. Conversely, planting earlier if favorable conditions are expected for germination and early growth.
- **Irrigation:** Adjusting irrigation schedules based on rainfall forecasts. If rain is expected, irrigation can be reduced or postponed, conserving water and energy. During dry spells, forecasts help determine when and how much to irrigate.
- **Fertilization:** Timing fertilizer application to coincide with favorable weather conditions for nutrient uptake and to avoid losses due to heavy rainfall runoff.
- **Pest and Disease Management:** Scheduling spraying or other control measures based on weather conditions that might favor the spread of pests or diseases (e.g., high humidity) or that might affect the efficacy of treatments (e.g., avoiding spraying before heavy rain).
- **Harvesting:** Planning harvest timing to coincide with dry weather to ensure optimal crop quality and reduce post-harvest losses. Farmers might expedite harvesting if a storm is predicted.
- **Frost Protection:** Implementing measures like covering plants or using wind machines when frost is forecast.
- **Livestock Management:** Providing shelter for animals during extreme heat or cold predicted by the forecast.

Essentially, utilizing weather forecasts allows farmers to be proactive rather than reactive, making more informed decisions that can lead to better outcomes. This requires access to reliable and timely weather information and the knowledge to interpret and apply it to specific agricultural practices.

1.2.2. Managing agricultural operations based on weather forecasts, particularly concerning extreme weather conditions

It involves taking proactive measures to protect crops, livestock, and infrastructure from potentially damaging events. This includes:

- **Preparing for Heat Waves:** Providing shade and increasing water availability for livestock. For crops, ensuring adequate irrigation and potentially using anti-transpirants.
- **Responding to Frost:** Implementing frost protection measures like covering sensitive plants, using wind machines, or applying protective sprays when freezing temperatures are forecast.
- **Preparing for Heavy Rainfall and Floods:** Ensuring proper drainage in fields, moving livestock to higher ground, and securing equipment. Delaying field operations like planting or fertilizer application if intense rainfall is expected.
- **Responding to High Winds and Storms:** Providing shelter for livestock, securing structures like greenhouses, and potentially harvesting mature crops early to prevent damage.
- **Managing During Droughts:** Implementing water conservation strategies, selecting drought-tolerant crops, and optimizing irrigation based on long-term precipitation forecasts.

The goal here is to use forecast information to anticipate extreme events and take timely actions to minimize their negative impact on agricultural productivity and assets.

1.3. Agricultural Meteorology

Agricultural Meteorology (also known as Agrometeorology) is an applied science that studies the relationship between **weather and climate** and **agriculture** in its broadest sense (Fig. 1.2).

Fig. 1.2: Weather, climate and agriculture

It investigates how meteorological and hydrological factors influence all aspects of agriculture, including:

- Crop production including horticulture, olericulture etc.
- Animal husbandry
- Forestry
- Pests and diseases
- Soil management

Essentially, it's the application of meteorological knowledge to agricultural practices to optimize production and minimize weather-related risks.

1.3.1. Scope of Agrometeorology

The scope of agricultural meteorology is quite broad and includes:

- **Studying crop-weather relationships:** Understanding how different weather parameters (temperature, rainfall, radiation, humidity, wind) affect the growth, development, and yield of various crops.
- **Analyzing agroclimatic resources:** Assessing the suitability of different regions for specific crops based on their climate.
- **Weather forecasting for agriculture:** Utilizing weather predictions (short-term, medium-term, and long-term) to aid in agricultural decision-making.
- **Pest and disease management:** Studying the influence of weather on the occurrence and spread of agricultural pests and diseases to develop effective control strategies.
- **Water management:** Applying meteorological data to optimize irrigation scheduling and water use efficiency.

- **Microclimate modification:** Investigating and implementing techniques to alter the immediate atmospheric environment around plants and animals to enhance their productivity.
- **Climate change impacts on agriculture:** Studying the effects of long-term climate variability and change on agricultural systems and developing adaptation and mitigation strategies.
- **Developing agro-advisory services:** Translating weather and climate information into practical advice for farmers.
- **Crop modeling:** Using meteorological data to simulate crop growth and predict yields.

1.3.2. Objectives of Agrometeorology

The primary objectives of agricultural meteorology are to:

- Understand and quantify the impact of weather and climate on agricultural production.
- Provide meteorological and related services to the agricultural community to support decision-making.
- Develop sustainable and economically viable agricultural systems.
- Improve agricultural production and quality.
- Reduce losses and risks associated with adverse weather conditions.
- Increase efficiency in the use of resources like water, labor, and energy in agriculture.
- Contribute to the development of climate-resilient agricultural practices.

1.3.3. Importance of Agrometeorology in Agriculture

Agricultural meteorology is of paramount importance because:

- **It directly influences crop growth and yields:** Weather conditions are fundamental drivers of plant development and productivity.
- **It aids in planning cropping systems and sowing dates:** Understanding climate helps in choosing suitable crops and optimal planting times.
- **It guides irrigation and fertilization practices:** Meteorological data informs when and how much to irrigate and can affect nutrient availability.
- **It helps in predicting and managing pests and diseases:** Weather conditions often favor the outbreak and spread of these problems.
- **It assists in mitigating the impacts of extreme weather events:** Forecasts allow farmers to take protective measures against frost, heat waves, floods, etc.

- **It contributes to efficient resource management:** By optimizing water and fertilizer use based on weather information.
- **It plays a crucial role in adapting to climate change:** Providing insights and strategies for dealing with altered weather patterns.

In essence, agricultural meteorology empowers farmers and agricultural stakeholders with the knowledge and tools to make informed, weather-smart decisions, leading to more efficient, productive, and sustainable agricultural practices.

2

Earth's Atmosphere, Weather and Climate

Earth's atmosphere is a dynamic and complex system that plays a crucial role in supporting life as we know it. This chapter explores into the fascinating world of this gaseous envelope that surrounds our planet, exploring its composition, structure, and the processes that drive both our daily weather and long-term climate. It includes:

- **Composition and Structure:** It will begin by examining the unique blend of gases that make up the atmosphere, including nitrogen, oxygen, argon, and trace amounts of others. It will also be explored how the atmosphere is structured into distinct layers (troposphere, stratosphere, mesosphere, thermosphere, and exosphere), each with its own characteristics and role.
- **Weather:** Weather refers to the short-term conditions of the atmosphere at a specific time and place. We'll discuss the key elements of weather, such as temperature, pressure, humidity, wind, and precipitation, and how they interact to create the ever-changing weather patterns we experience.
- **Climate:** In contrast to weather, climate describes the long-term average of atmospheric conditions in a particular region. We'll investigate the factors that influence climate, including latitude, altitude, proximity to oceans, and atmospheric circulation patterns. We'll also explore the different climate zones found across the globe.
- **Interconnectedness:** A key focus of this chapter understands how weather and climate are interconnected. Daily weather patterns are influenced by the same fundamental atmospheric processes that shape long-term climate. For example, the way the sun's energy interacts with the atmosphere drives both daily temperature variations and global climate patterns.

2.1 Importance of the Atmosphere for Agriculture and Food Production

The atmosphere plays an absolutely critical role in agriculture and global food production. Here's how:

- **Temperature Regulation:** The atmosphere's greenhouse effect keeps Earth warm enough to support plant growth. Different plants have optimal temperature ranges for growth, and the atmosphere helps maintain these conditions.
- **Water Cycle:** The atmosphere drives the water cycle, which provides essential rainfall for crops. Evaporation, condensation, and precipitation are all atmospheric processes.
- **Air Composition:** Plants need carbon dioxide for photosynthesis, a process that converts sunlight into energy for growth. The atmosphere provides this vital gas.
- **Protection from Harmful Radiation:** The ozone layer in the stratosphere absorbs harmful ultraviolet (UV) radiation from the sun, which can damage plant tissues and hinder growth.
- **Weather Patterns:** Predictable weather patterns, influenced by atmospheric conditions, are essential for agricultural planning, planting, and harvesting.
- **Climate and Growing Seasons:** Climate determines which crops can be grown in a particular region and the length of the growing season.

Any significant changes in the atmosphere, such as those seen with climate change, can have profound impacts on agriculture and food production, threatening global food security.

By the end of this chapter, you will gain a deeper appreciation for the delicate balance of Earth's atmospheric system and the forces that shape our ever-changing world.

2.1.1. Earth's Atmosphere: A Life-Sustaining Shield

Earth's atmosphere is a complex, dynamic, and life-sustaining layer of gases that surrounds our planet. Held in place by Earth's gravity, it extends from the surface to space, gradually thinning until it becomes negligible. This gaseous envelope is essential for regulating temperature, enabling weather patterns, and protecting life from harmful radiation.

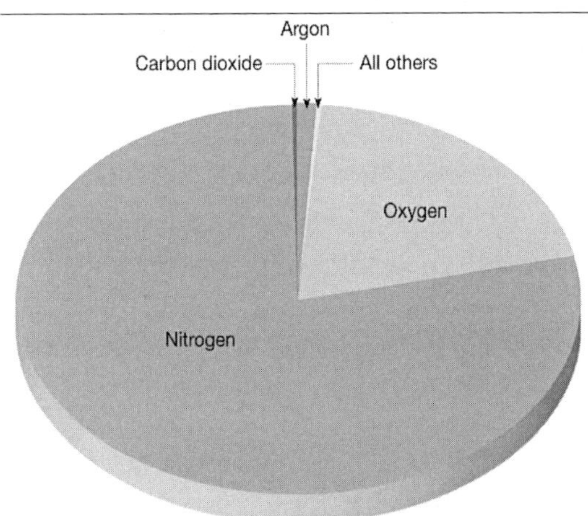

Fig. 2.1: Composition of Earth's atmosphere

- **Composition :** The composition of Earth's atmosphere is a delicate balance of several gases, with regional and temporal variations. The major components of dry air are:
- **Nitrogen (N_2):** Approximately 78%. Nitrogen is relatively inert and plays a crucial role in diluting oxygen, preventing rapid combustion.
- **Oxygen (O_2):** Approximately 21%. Oxygen is vital for respiration in most living organisms and is essential for combustion. It's primarily produced by photosynthesis.
- **Argon (Ar):** Approximately 0.9%. Argon is an inert noble gas.

In addition to these major components, the atmosphere contains several variable components that play critical roles:

- **Water Vapor (H_2O):** Its concentration varies widely (0-4%) depending on location and time. Water vapor is crucial for humidity, cloud formation, and precipitation.
- **Carbon Dioxide (CO_2):** A trace gas (around 0.04% or 425 parts per million currently), but its concentration is increasing due to human activities. CO_2 is a greenhouse gas, playing a vital role in regulating Earth's temperature and is crucial for photosynthesis.
- **Methane (CH_4):** Another potent greenhouse gas, present in small amounts. Methane is emitted from natural sources (wetlands, decomposition) and human activities (agriculture, fossil fuel production).
- **Ozone (O_3):** Concentrated in the stratosphere, the ozone layer absorbs most of the Sun's harmful ultraviolet (UV) radiation.

- **Aerosols:** Tiny solid and liquid particles (dust, sea salt, volcanic ash, pollution) that influence cloud formation and affect the amount of sunlight that reaches the surface.

2.2. Structure of the Atmosphere

The vertical structure of the atmosphere is characterized by distinct layers, each with its own temperature profile, composition, and phenomena (Fig. 2.2).

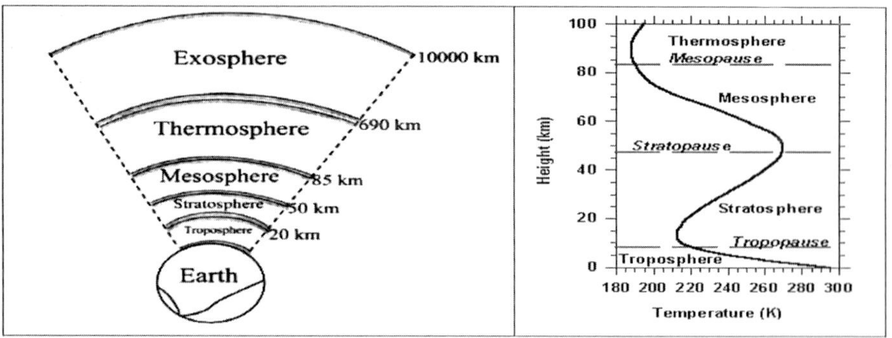

Fig. 2.2: Vertical structure of the atmosphere

These layers, from the Earth's surface upwards, are:

2.2.1. Troposphere

- **Altitude:** Extends from the surface to an average of about 10-15 km (The height of troposphere is maximum near the equator and decreases with increase in latitude with minimum height near the poles)
- **Temperature:** Generally decreases with altitude at an average rate of 6.5°C per kilometer (known as the lapse rate).
- **Characteristics**
  - Contains about 75% of the atmosphere's mass and almost all of its water vapor.
  - This is where most weather phenomena (clouds, rain, and snow) occur due to the presence of water vapor and air movement.
  - Heated from the ground up, as the Earth's surface absorbs solar radiation and transfers heat to the air above.
  - The top of the troposphere is called the **tropopause**.

2.2.2. Stratosphere

- **Altitude:** Extends from the tropopause to about 50 km.
- **Temperature:** Increases with altitude.
- **Characteristics**

- Contains the **ozone layer** (around 15-35 km), which absorbs much of the Sun's harmful ultraviolet (UV) radiation, causing the temperature increase.
- Air is stable and less turbulent compared to the troposphere, which is why commercial jets often fly in the lower stratosphere.
- The top of the stratosphere is called the **stratopause**.

2.2.3. Mesosphere

- **Altitude:** Extends from the stratopause to about 80-85 km.
- **Temperature:** Decreases with altitude, reaching the coldest temperatures in the atmosphere (around −90°C) at the top.
- **Characteristics**
  - Air is very thin.
  - Most meteors burn up in this layer due to friction with the air molecules.
  - The top of the mesosphere is called the **mesopause**.

2.2.4. Thermosphere

- **Altitude:** Extends from the mesopause upwards to about 500-1,000 km.
- **Temperature:** Increases significantly with altitude due to the absorption of high-energy solar radiation (UV and X-rays). Temperatures can reach hundreds or even thousands of degrees Celsius.
- **Characteristics**
  - Despite the high temperatures, it would feel very cold because the air density is extremely low, meaning there are very few molecules to transfer heat.
  - The **ionosphere**, a region where gases are ionized by solar radiation, is located within the thermosphere (and extends into the lower exosphere). The ionosphere plays a crucial role in radio wave propagation and is where auroras (Northern and Southern Lights) occur.
  - The International Space Station orbits in the thermosphere.
  - The upper boundary is sometimes called the **thermopause** or **exobase**.

2.2.5. Exosphere

- **Altitude:** Extends from the thermopause outwards, with no well-defined upper limit, eventually merging into interplanetary space (can extend up to 10,000 km or more).

- **Temperature:** Extremely high in the lower parts, but the concept of temperature becomes less meaningful due to the very low density.
- **Characteristics**
  - The air is extremely thin, and molecules can escape into space.
  - Satellites also orbit in this region.

In summary, the vertical structure of the atmosphere is layered, with each layer exhibiting distinct temperature trends and playing a unique role in Earth's system. These layers are primarily defined by their temperature profiles, which are influenced by factors like solar radiation absorption and heat transfer mechanisms. The composition of the atmosphere is relatively uniform in the lower layers (homosphere, including the troposphere, stratosphere, and mesosphere) and becomes stratified by molecular weight in the upper layers (heterosphere, mainly the thermosphere and exosphere).

2.3. Weather

- Weather refers to the **state of the atmosphere at a particular time and place**. It describes the current or very short-term (from minutes to days) conditions of the atmosphere with respect to variables such as temperature, pressure, humidity, wind, and precipitation. Essentially, it's what's happening outdoors right now or what is expected to happen in the near future.
- Weather is dynamic and constantly changing due to the complex interactions of various atmospheric elements. It's important to distinguish weather from **climate**, which describes the average weather conditions of a region over a long period (typically 30 years or more). Think of weather as your daily mood, while climate is your overall personality.

2.3.1. Weather Elements

These are the measurable quantities or phenomena that describe the state of the atmosphere at any given time. The primary elements of weather include:

- **Temperature:** How hot or cold the air is. Measured using thermometers (in Celsius, Fahrenheit, or Kelvin).
- **Atmospheric Pressure:** The weight of the air pressing down on a unit area. Measured using barometers (in hectopascals or millibars). Changes in pressure are often associated with changes in weather.
- **Wind:** The movement of air, described by its speed and direction. Wind speed is measured by anemometers (in kilometers per hour, miles per hour, or knots), and wind direction is indicated by wind vanes or weathercocks.

- **Humidity:** The amount of water vapor present in the air. It can be expressed as absolute humidity (mass of water vapor per unit volume of air) or relative humidity (percentage of the maximum amount of water vapor the air can hold at a given temperature). Measured using hygrometers.
- **Precipitation:** Any form of water that falls from clouds to the Earth's surface. This includes rain, snow, sleet, and hail. Measured using rain gauges (in millimeters or inches).
- **Clouds:** Visible masses of water droplets or ice crystals suspended in the atmosphere. They are classified by their form, altitude, and whether they produce precipitation. Cloud cover is often described in oktas (eighths) of the sky covered.
- **Visibility:** The distance to which an object can be clearly seen. Reduced visibility can be caused by fog, haze, dust, or heavy precipitation. Measured using visibility sensors.
- **Sunshine Duration:** The amount of time the Earth's surface is directly exposed to solar radiation. Measured in hours.

These elements are interconnected, and their interplay determines the type of weather we experience.

2.3.2. Weather Characteristics

These are the general features or behaviors of weather that we observe and experience:

- **Dynamic and Changeable:** Weather is in constant flux. Conditions can change rapidly, from sunny to rainy in a short period.
- **Localized:** Weather conditions can vary significantly over relatively small distances. For example, it might be raining in one part of a city and sunny in another.
- **Influenced by Multiple Factors:** Weather is driven by a complex interplay of solar radiation, the Earth's rotation, the distribution of land and water, topography, and global air and ocean currents.
- **Described by Extremes:** We often characterize weather by its extremes, such as record high or low temperatures, the heaviest rainfall, or the strongest winds.
- **Predictable (to a degree):** While inherently complex, weather patterns can be predicted using scientific models and observations, allowing for weather forecasting. However, the accuracy of forecasts generally decreases with longer time horizons.

- **Impactful:** Weather significantly affects our daily lives, from what we wear to our travel plans, and can have major impacts on agriculture, infrastructure, and safety.
- In essence, weather is the ever-changing atmospheric story of our planet on a short timescale, shaped by the continuous interaction of its fundamental elements.

2.4. Atmospheric Weather Variables

These are the fundamental properties of the atmosphere that are measured and observed to describe the weather at any given time.

- **Temperature**
  - Measures how hot or cold the air is.
  - Typically measured in Celsius (°C), Fahrenheit (°F), or Kelvin (K) using thermometers.
  - Varies with altitude (generally decreases in the troposphere), latitude, time of day, and season.
  - Influences other weather elements like humidity and precipitation.
- **Atmospheric Pressure**
  - The force exerted by the weight of the air above a given point.
  - Measured using barometers in units like hectopascals (hPa) or millibars (mb).
  - Decreases with altitude.
  - Horizontal differences in pressure drive the wind. High-pressure systems are often associated with fair weather, while low-pressure systems can bring stormy conditions.
- **Humidity**
  - The amount of water vapor in the air.
  - Can be expressed in several ways:
    - **Absolute humidity:** The mass of water vapor per unit volume of air.
    - **Specific humidity:** The mass of water vapor per unit mass of air.
    - **Relative humidity (RH):** The percentage of the maximum amount of water vapor the air can hold at a given temperature. This is what is commonly reported.
    - **Dew point:** The temperature to which air must be cooled at a constant pressure and constant water vapor content for saturation to occur.A high dew point indicates a lot of moisture in the air.

Measured using hygrometers, humidity plays a crucial role in cloud formation, precipitation, and how we perceive temperature.

- **Wind**
  - The horizontal movement of air.
  - Described by its **speed** (measured by anemometers in units like km/h, mph, or knots) and **direction** (indicated by wind vanes).Wind direction is usually reported as where the wind is blowing from.
  - Caused by differences in air pressure. Air moves from areas of high pressure to areas of low pressure. The greater the pressure difference, the stronger the wind.
  - Global wind patterns (like trade winds and westerlies) and local winds (like sea breezes and land breezes) are important aspects of weather.
- **Precipitation**
  - Any form of water that falls from clouds to the Earth's surface.
  - Includes rain, snow, sleet, and hail.
  - Measured using rain gauges (in mm or inches) or by snow depth.
  - Forms when water vapor in the atmosphere condenses or freezes and becomes heavy enough to fall.
- **Clouds**
  - Visible masses of water droplets or ice crystals suspended in the atmosphere.
  - Classified by their altitude, shape, and whether they produce precipitation (e.g., cumulus, stratus, cirrus, nimbus).
  - Cloud cover is often described as the fraction of the sky obscured by clouds (e.g., clear, partly cloudy, overcast).
- **Visibility**
  - The distance to which an object can be clearly seen.
  - Reduced by fog, haze, smoke, dust, and heavy precipitation.
  - Important for transportation safety (aviation, driving, shipping).

These atmospheric variables are interconnected and constantly interacting, leading to the diverse weather conditions we experience. Meteorologists observe and measure these variables to understand the current state of the atmosphere and to forecast future weather.

2.5 Climate

Climate, in simple terms, is the **long-term average of weather conditions** in a particular region. It encompasses the typical weather patterns, the frequency and intensity of extreme weather events, and the variations from year to year over an extended period, usually 30 years or more as defined by the World Meteorological Organization (WMO).

Think of it this way: **Weather** tells you what clothes to wear today, while **climate** tells you what kinds of clothes you generally need in your wardrobe for the whole year in a specific location.

Climate is described by statistical information such as average temperature, precipitation, humidity, wind, and sunshine over a long period. It also includes information about the range and variability of these conditions.

2.5.1. Factors Influencing Climate

The climate of a region is influenced by a variety of interacting factors, both natural and human-induced:

- **Latitude**
  - The distance from the equator. Regions closer to the equator receive more direct sunlight and are generally warmer, while regions closer to the poles receive less direct sunlight and are colder.
  - Latitude also affects the length of day and night, which influences temperature variations throughout the year.
- **Altitude (Elevation)**
  - The height above sea level. Temperature generally decreases with increasing altitude because the air is less dense and retains less heat.
- **Distance from the Sea (Continentality)**
  - Large bodies of water like oceans and large lakes moderate temperatures. Coastal areas tend to have milder winters and cooler summers compared to inland areas at the same latitude. Water heats up and cools down more slowly than land.
- **Ocean Currents**
  - The movement of ocean water transports heat around the globe. Warm currents can warm coastal regions that would otherwise be colder, and cold currents can cool otherwise warmer areas. For example, the Gulf Stream warms Western Europe.
- **Prevailing Winds**
  - The dominant direction of wind in a region affects temperature and precipitation. Winds blowing from the sea can bring moisture, while winds from large landmasses can be dry.
- **Topography (Relief)**
  - The shape of the land, such as mountains, can significantly impact climate. Mountains can cause orographic precipitation (where air is

forced to rise, cool, and condense, leading to rain or snow on the windward side, while the leeward side becomes drier, creating a rain shadow). Altitude also plays a role here (as mentioned above).

- **Vegetation**
    - Plant cover can influence local temperature and humidity through processes like evapotranspiration (the release of water vapor from plants and soil). Forests, for example, can lead to more local rainfall.
- **Human Activities**
    - The burning of fossil fuels, deforestation, and industrial processes have increased the concentration of greenhouse gases in the atmosphere, leading to global warming and climate change. Changes in land use also affect local and regional climates.

2.5.2. Characteristics of Climate

Climate can be characterized by several aspects:

- **Long-Term Averages:** The most fundamental characteristic is the average values of weather elements (temperature, precipitation, etc.) calculated over a long period.
- **Variability:** Climate descriptions also include the typical range of conditions and how much they vary from year to year or season to season. Some climates have high variability, while others are more consistent.
- **Extremes:** The frequency and intensity of extreme weather events (heatwaves, cold spells, droughts, floods, storms) are important characteristics of a region's climate.
- **Distinct Seasons:** Many regions have climates characterized by distinct seasons with noticeable changes in temperature, precipitation, and daylight hours.
- **Spatial Patterns:** Climate varies geographically, resulting in different climate zones across the Earth (e.g., tropical, temperate, polar, arid). These zones have characteristic temperature and precipitation regimes.
- **Change over Time:** While traditionally viewed as relatively stable over human timescales, we now know that climate can change over time due to both natural factors and human activities. This change is a significant characteristic of the current global climate.

In summary, climate provides a comprehensive picture of the typical atmospheric conditions of a place over the long term, shaped by a multitude of interacting factors and described by its averages, variability, extremes, and changes.

2.6. Differences between Weather and Climate

The weather and climate can be distinguished based on the differences as explained in the table given below:

Feature	Weather	Climate
What it is	The **here and now** condition of the atmosphere. Think about what you experience when you step outside *today*. Is it hot, cold, windy, or raining? Weather is like a snapshot or a brief video of what's happening in the sky right now.	The **usual** or **typical** weather pattern of a place **over a long time**. It's the overall picture built up from many years of weather observations. Climate tells you what kind of weather you can generally expect in a region throughout the year.
Time Focus	**Short-term**. We talk about the weather for today, tomorrow, or maybe the next week. It's about what's happening in a limited timeframe.	**Long-term**. Climate is about averages and patterns that have been observed over many decades (usually at least 30 years). It looks at the big picture over a significant period.
What it describes	The specific values of atmospheric elements like temperature (e.g., 30°C as its currently early morning in Delhi), precipitation (e.g., likely dry right now), wind, humidity, and cloud cover *at* this moment or in the near future.	The typical ranges and averages of these atmospheric elements over many years. For example, the average summer temperature, the average annual rainfall, the usual wind patterns for a season in a particular location. It also includes how often extreme events (like heatwaves or droughts) tend to occur.
How it changes	Can change very rapidly – from sunshine to rain in an hour, or from warm to cool overnight. It's dynamic and constantly fluctuating.	Changes more slowly. While there can be warmer or cooler years, the underlying climate of a region tends to be relatively stable unless influenced by significant factors like natural climate variability or human-caused climate change (which occur over longer periods).
Pre-diction	We use weather forecasts to predict what the weather will be like in the coming hours or days. These forecasts become less accurate the further into the future they models are used to predict long-term trends and changes in climate, such as how average tempe ratures or rainfall patterns might shift over decades due to greenhouse gas emissions. These are about long-term probabilities and ten-dencies, not specific day-to-day condi-tions far in the future.	The trends in change of climate can only be predicted.

Example	„Currently in Delhi, it‘s likely mild with clear skies.“ (A general statement fitting the early morning)	„The climate of Delhi is generally characterized by hot, dry summers and a monsoon season with moderate rainfall from June to September.“

Thus weather and climate can be distinguished from each other as given below

- **Timescale**
  - **Weather:** Describes atmospheric conditions over a **short period** (minutes, hours, days, weeks).
  - **Climate:** Describes the average atmospheric conditions over a **long period** (typically 30 years or more).
- **Variability**
  - **Weather:** Is **highly variable** and can change rapidly.
  - **Climate:** Is more **stable**, representing long-term patterns. Changes in climate occur over much longer timescales.
- **Focus**
  - **Weather:** Focuses on the **specific conditions** of the atmosphere at a given time (e.g., the temperature right now, whether it's raining).
  - **Climate:** Focuses on the **typical or average conditions** of a region over many years (e.g., the average temperature in summer, the usual amount of rainfall).
- **Predictability**
  - **Weather:** Can be predicted in the short term using weather forecasts, but accuracy decreases with longer time horizons.
  - **Climate:** Long-term trends and averages can be predicted using climate models, focusing on overall changes over decades or centuries.
- **Geo/Spatial Scale**
  - **Weather:** Can be much localized (e.g., raining in one part of a city but not another).
  - **Climate:** Typically describes the conditions of a larger region or even the entire globe.

In essence, think of weather as the day-to-day state of the atmosphere, while climate is the long-term summary of that weather (Fig. 2.3).

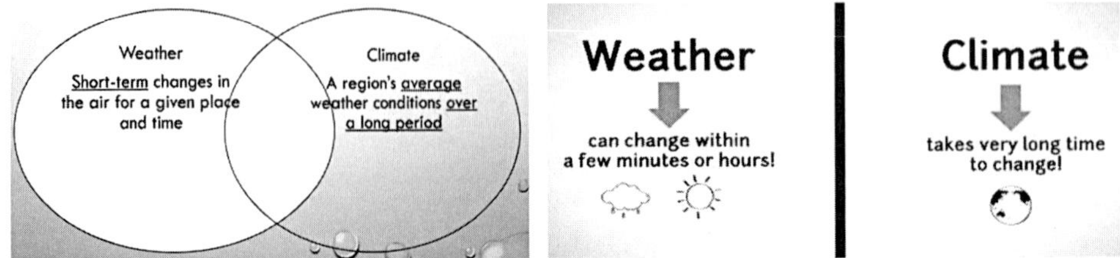

Fig. 2.3: Weather and climate

2.7. Geo-coordinates and Time Conversion

2.7.1. Geo-coordinates

Also known as geographic coordinates, are numerical values used to precisely locate a point on the Earth's surface (Fig. 2.4). They are composed of latitude and longitude, which define a location's position relative to the equator and prime meridian.

- **Latitudes** are horizontal lines that measure the distance north or south of the Equator, ranging from 0° at the Equator to 90° at the poles.
- **Longitudes** are vertical lines, measuring the distance east or west of the Prime Meridian, which is 0° longitude and runs through **Greenwich, England**. They have one very important function; they determine local time in relation to GMT or Greenwich Mean Time, which is sometimes referred to as **World Time**.
- **Altitude**, is the distance above sea level. Areas are often considered "high-altitude" if they reach at least 2,400 meters (8,000 feet) into the atmosphere. The most high-altitude point on Earth is Mount Everest, in the Himalayan mountain range on the border of Nepal and the Chinese region of Tibet. Mount Everest is 8,850 meters (29,035 feet) tall.

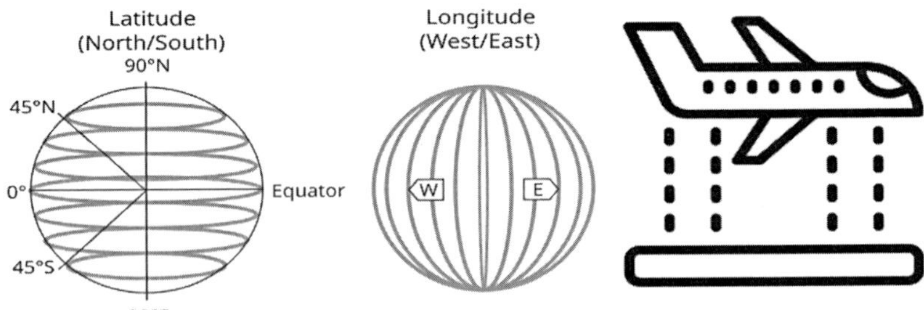

Fig. 2.4: Geo-coordinates

Since the earth makes one complete revolution of 360° in one day or 24 hours, it passes through 15° in one hour or 1° in 4 minutes. The earth rotates from west to east, so every 15° we go eastwards, local time is advanced by 1 hour. Conversely, if we go westwards, local time is retarded by 1 hour. It may be concluded that places east of Greenwich see the sun earlier and gain time, whereas places west of Greenwich see the sun later and lose time.

- If we know GMT, to find local time, we merely have to add or subtract the difference in the number of hours from the given longitude.
- Indian Standard Time (IST) is determined based on the 82.5° East meridian, which passes near Allahabad/Prayagraj in Uttar Pradesh. This meridian serves as the standard time reference for the entire country.
- To convert IST (Indian Standard Time) to LMT (Local Mean Time) for weather observations in India, we need to know the longitude of the observation location, because LMT is based on the local longitude, whereas IST is fixed at 82.5°E longitude.

2.7.2. Formula to Convert IST to LMT

LMT = IST − (Longitude − 82.5) × 4 minutes

The Earth rotates **1° every 4 minutes**.

IST is the mean time of **82.5°E longitude** (standard meridian of India).

If your location is **east** of 82.5°, LMT will be **ahead** of IST.

If your location is **west** of 82.5°, LMT will be **behind** IST.

- ***Example 1: For Delhi (Longitude ≈ 77.1°E)***

  Difference= 77.1 − 82.5 = −5.4°

  Time difference= −5.4 × 4 = −21.6 minutes

  LMT=IST−21.6 minutes

  So, LMT in **Delhi is about 21.6 minutes behind IST**.

- ***Example 2: For Guwahati (Longitude ≈ 91.7°E)***

  Difference= 91.7 − 82.5 = 9.2°

  Time difference=9.2×4=36.8 minutes

  LMT = IST + 36.8 minutes

  So, LMT in Guwahati is about 36.8 minutes ahead of IST.

3

Atmospheric Pressure and Wind

Imagine the air around us. Even though we can't see it, air has weight. Atmospheric pressure is essentially the **force exerted by the weight of the air molecules above a particular point on the Earth's surface.**

Think of it like this: Imagine you're at the bottom of a swimming pool. The water above presses you down, and you feel that pressure. Similarly, the atmosphere, a giant ocean of air, has layers of air molecules, and their weight presses down on everything below, including us.

- **Atmospheric pressure** is defined as the **force per unit area exerted by the weight of the air column extending from the surface of the earth to the top of the atmosphere above that area** (Fig. 3.1).

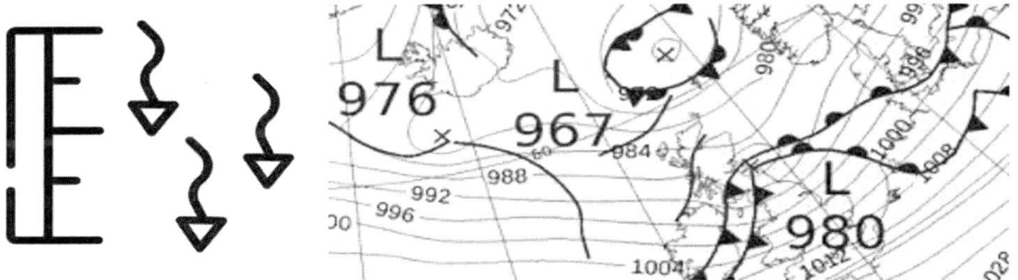

Fig. 3.1: Atmospheric pressure

- It can be explain in simple terms like
  - **Air has weight:** Just like water or a pile of books, the air surrounding the Earth has mass, and thus, weight due to gravity.
  - **Weight creates pressure:** This weight of the air presses down on the surface.6 The more air there is above a certain point, the greater the pressure.
  - **Like a stack:** Imagine a stack of blankets. The blanket at the bottom has to support the weight of all the blankets above it, so it experiences more pressure. Similarly, the air at lower altitudes has the weight of all the air above pressing down on it.

The most common formula used to calculate atmospheric pressure, particularly when considering a column of fluid is: $P = \rho gh$, Where:

- **P** = Atmospheric pressure. This is the force per unit area exerted by the weight of the air. It's typically measured in Pascals (Pa), but other units like atmospheres (atm), millibars (mb), or inches of mercury (inHg) are also used.
- ρ (rho) = Density of the fluid. In the case of atmospheric pressure measured by a barometer, the fluid is often mercury, so you'd use the density of mercury. For calculations involving air itself, you'd use the density of air. Density is measured in kilograms per cubic meter (kg/m^3).
- **g** = Acceleration due to gravity. This is the constant acceleration of an object falling freely near the Earth's surface. Its standard value is approximately 9.81 meters per second squared (m/s^2).
- **h** = Height of the fluid column. This is the vertical height of the column of fluid (mercury, in a barometer) that is being supported by the atmospheric pressure. It's measured in meters (m).

In simpler terms, this formula tells us that the pressure at the bottom of a column of fluid is directly proportional to the density of the fluid, the acceleration due to gravity, and the height of the column.

3.1. Measurement of Atmospheric Pressure

The instrument used to measure atmospheric pressure is called Barometer. The Fortin barometer is a type of mercury barometer used for precise measurement of atmospheric pressure. It consists of a narrow **vertically mounted** glass tube, approximately 90 cm in length, enclosed within a brass tube for protection. The glass tube is filled with mercury, and the lower end is immersed in a cistern containing mercury (Fig. 3.2).

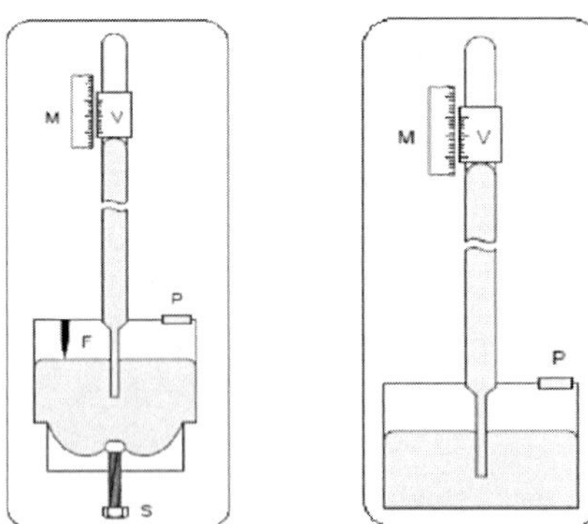

Fig. 3.2: Fortin barometer to measure atmospheric pressure

- **Key components of barometer include**
  - **Glass tube:** Contains the mercury column and is mounted vertically.
  - **Brass tube:** Protects the glass tube and has a slit to view the mercury level.
  - **Cistern:** A reservoir at the bottom containing mercury (P). The bottom is flexible (leather) and can be adjusted.
  - **Ivory pointer:** Located inside the cistern, its tip marks the zero point of the scale (F)
  - **Main scale:** A graduated scale in millimeters attached to the brass tube (M)
  - **Vernier scale:** A sliding scale for precise measurements (V)
  - **Adjusting screw:** Used to raise or lower the mercury level in the cistern to align with the ivory pointer.(S)
- **To take a reading**
  - Use the adjusting screw to make the mercury level in the cistern just touch the tip of the ivory pointer.
  - Observe the top of the mercury column in the glass tube.
  - Adjust the vernier scale so its edge is tangent to the meniscus (the curved surface of the mercury).
  - Read the main scale and the vernier scale to get the atmospheric pressure.

The height of the mercury column from the top of the pointer to the level of mercury in the glass tube denotes the atmospheric pressure in Cm of Hg.

Common units to express atmospheric pressure include

- **Pascals (Pa) or Hectopascals (hPa):** The standard SI unit.
- **Millibars (mb):** Often used in meteorology (1 hPa = 1 mb).
- **Inches of mercury (inHg) or millimeters of mercury (mmHg):** Historically used in barometers.
- **Atmospheres (atm):** Roughly the average pressure at sea level.

The average atmospheric pressure at sea level is about **1013.25 hPa** (or mb), which is also equal to **1 atm**.

- **About pascal, bar, and millibar**
  - **Pascal (Pa):** The pascal is the SI unit of pressure. It is defined as one newton per square meter (N/m^2). In base SI units, this is equivalent to one kilogram per meter per second squared ($kg.m^{-1}.s^{-2}$). The pascal is a relatively small unit, so kilopascals (kPa) are often used.

- **Bar (bar):** The bar is a metric unit of pressure defined as exactly 100,000 pascals (105 Pa). It is slightly less than the standard atmospheric pressure.
- **Millibar (mbar):** The millibar is a unit of pressure equal to one-thousandth of a bar (10−3 bar). Therefore, 1 mbar=100 Pa. Millibars are commonly used in meteorology to measure atmospheric pressure.

The **newton (N)** is the standard unit of force in the International System of Units (SI). One newton is defined as the force required accelerating a mass of one kilogram (1 kg) at a rate of one meter per second squared (1 m/s^2). Mathematically, this is expressed as: 1 N=1 $kg.m/s^2$

In simpler terms, if a force of one newton to an object with a mass of one kilogram, the object will accelerate at one meter per second every second.

The newton is named after **Sir Isaac Newton** in recognition of his work on classical mechanics, particularly his laws of motion.

3.1.1. Conversion of Atmospheric Pressure in various Units

We know that 1 atm=76 Cm of Hg.

From the table given below, the relationship between Pascals and Newtons can be noted:

Unit	**Abbreviation**	**Equivalent to 1 atm (approx)**	**Notes**
Pascal	Pa	101325 Pa	1 Pa=1 N/m^2
Kilopascal	kPa	101.325 kPa	
Bar	bar	1.01325 bar	
Millibar	mbar	1013.25 mbar	
Torr	Torr	760 Torr	
Millimeters of mercury	mmHg	760 mmHg	
Centimeters of mercury	cmHg	76 cmHg	
Pounds per square inch	psi	14.696 psi	
Inches of mercury	inHg	29.92 inHg	

3.2. Standard Atmospheric Pressure

Standard Atmospheric Pressure (atm) is defined as the average atmospheric pressure at sea level. It's a standard reference point for expressing pressure and is used in various scientific and technical contexts.

The current official definition of the standard atmosphere, as adopted by the 10th General Conference on Weights and Measures (CGPM) in 1954, is:

1 atm=101325 Pascals (Pa)

This value is also equivalent to:

- 760 millimeters of mercury (mmHg)
- 76 centimeters of mercury (cmHg)
- 29.92 inches of mercury (inHg)
- 1013.25 millibars (mbar)
- 1.01325 bar
- 14.696 pounds per square inch (psi)

Essentially, standard atmospheric pressure is the pressure exerted by the Earth's atmosphere at mean sea level under specific conditions.

3.2.1. Variation of Atmospheric Pressure with Increase in Height

Atmospheric pressure at any point is essentially the **weight of the column of air above that point**. It may be explained as:

- **At sea level**, you have the entire atmosphere pressing down on you. The pressure you measure is due to the weight of all that air.
- As you move to a **higher altitude** (like a mountaintop), there is less air above you. Consequently, the weight of the air column pressing down is less.
- Therefore, the atmospheric pressure decreases with increasing height because there is less air above to exert that downward force.

The decrease in the weight of the air column is directly related to two main factors as we ascend:

- **Fewer Air Molecules:** The higher we go, the fewer air molecules there are in a given volume. Gravity pulls most of the air molecules closer to the Earth's surface. This means the **number density** of air molecules decreases with altitude.
- **Decreased Air Density:** Because there are fewer molecules in the same amount of space, the **density of the air** also decreases with altitude. Less dense air weighs less per unit volume.

3.2.2. Comparison of Atmospheric Pressure Measured at various Locations

To compare atmospheric pressure measured at different places accurately, we need to account for the factors that naturally cause pressure variations. The most significant of these is **altitude**.

Typically atmospheric pressure may be compared as

- **Reduce to Sea Level:** To eliminate the effect of altitude, atmospheric pressure readings are often converted or "reduced" to what the pressure would be if the measurement were taken at sea level. This standardized value allows for meaningful comparisons between locations at different elevations. Meteorological organizations often use standard formulas that take into account the station's elevation and sometimes the local air temperature to perform this reduction. The resulting pressure is often referred to as the **Mean Sea Level Pressure (MSLP)**.

- **Use the Same Units:** Ensure that all pressure readings being compared are in the same units (e.g., hectopascals, inches of mercury). If they are in different units, you'll need to convert them using the conversion factors we discussed earlier.
- **Consider Other Factors:** While reducing to sea level largely addresses altitude differences, other factors can cause real variations in atmospheric pressure between locations at similar (or even the same) altitudes. These include:
    - **Temperature:** Warmer air is generally less dense and exerts lower pressure, while cooler air is denser and exerts higher pressure.
    - **Humidity:** Moist air is slightly less dense than dry air at the same temperature and pressure, so higher humidity can lead to slightly lower pressure.
    - **Weather Systems:** High-pressure systems and low-pressure systems cause significant regional variations in atmospheric pressure.
- **In practice, when you see weather maps with isobars (lines connecting points of equal pressure), these pressures have usually been reduced to sea level.** This allows meteorologists to easily identify high and low-pressure areas, which are crucial for weather forecasting.
- So, the primary method for comparing atmospheric pressure at different places is to **reduce the measurements to sea level equivalent pressures** and then compare those values.
- The procedure to reduce atmospheric pressure measured at a **specific location to mean sea level** (MSL) involves using a formula that accounts for the elevation difference between the measurement site and sea level, and often the local air temperature. The goal is to estimate what the atmospheric pressure would be if the measurement were taken at sea level.

3.3. Wind

Air in horizontal motion is called wind. The direction **from** which the wind is coming is called **windward** direction. The direction **towards** which wind is blowing is called **leeward** direction (Fig. 3.3).

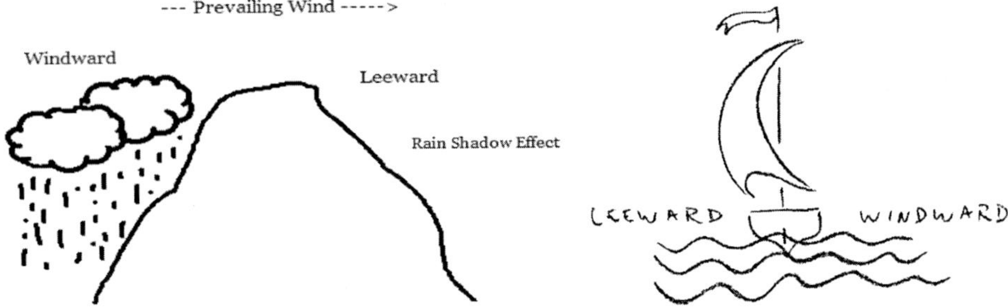

Fig. 3.3: Windward and leeward dirctions

3.3.1. Types of wind

Based on their nature and scale, winds can be broadly classified into three main types:

- **Permanent Winds (Planetary Winds):** These winds blow consistently throughout the year in a particular direction. They cover large areas of the globe. Examples include:
  - **Trade Winds:** Blow from the subtropical high-pressure belts towards the equatorial low-pressure belt. In the Northern Hemisphere,1 they are northeasterly, and in the Southern Hemisphere, they are southeasterly due to the Coriolis effect.
  - **Westerlies:** Blow from the subtropical high-pressure belts towards the subpolar low-pressure belts in the mid-latitudes. They are southwesterly in the Northern Hemisphere and northwesterly in the Southern Hemisphere.
  - **Polar Easterlies:** Blow from the polar high-pressure areas towards the subpolar low-pressure belts. They are northeasterly in the Northern Hemisphere and southeasterly in the Southern Hemisphere.
- **Seasonal Winds (Periodic Winds):** These winds change their direction with the change of seasons. The most prominent example is:
  - **Monsoon Winds:** These are large-scale seasonal winds that reverse their direction between summer and winter. They are particularly significant in South Asia.
- **Local Winds (Tertiary Winds):** These winds blow over smaller areas and are influenced by local temperature and pressure differences. They occur during particular times of the day or year. Examples include:
  - **Land and Sea Breezes:** Occur near coastlines due to the different heating and cooling rates of land and water. Sea breezes blow from

sea to land during the day, and land breezes blow from land to sea at night.

- **Mountain and Valley Breezes:** Occur in mountainous regions. Valley breezes blow uphill during the day as the mountain slopes heat up, and mountain breezes blow downhill at night as the slopes cool.
- **Specific Local Winds:** These have regional names, such as the Loo (hot, dry wind in Northern India and Pakistan), Chinook (warm, dry wind on the eastern slopes of the Rockies), Mistral (cold, northwesterly wind in southern France), etc.

3.4. Global Pressure and Wind Systems

Global pressure belts and wind systems are large-scale patterns of atmospheric circulation that shape our climate and weather. Let's break it down:

3.4.1. Global Pressure Belts

The Earth's atmosphere is divided into several pressure belts, which are areas of high or low pressure that encircle the globe (Fig. 3.4).

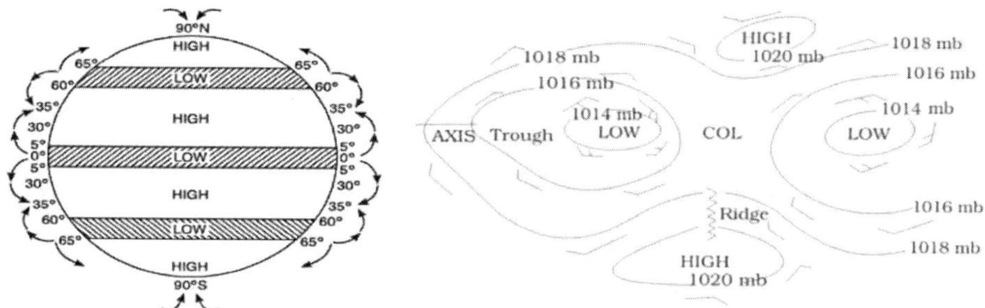

Fig. 3.4: Atmospheric pressure belts and wind systems

These belts are created by the uneven heating of the Earth's surface by the sun.

- **Equatorial Low-Pressure Belt**: Near the equator, the sun's rays strike the Earth directly, heating the air and causing it to rise. This creates a belt of low pressure near the equator.
- **Subtropical High-Pressure Belts**: At around 30°N and 30°S latitude, the air cools and sinks, forming high-pressure belts. These belts are also known as the horse latitudes.
- **Subpolar Low-Pressure Belts**: At around 60°N and 60°S latitude, the air cools and rises, creating low-pressure belts.
- **Polar High-Pressure Belts**: At the poles, the air is cold and dense, forming high-pressure belts.

3.4.2. Global Wind Systems

Wind is the movement of air from high-pressure areas to low-pressure areas. The global wind systems are driven by the pressure differences between the pressure belts.

- **Trade Winds**: Blow from the subtropical high-pressure belts towards the equatorial low-pressure belt. In the Northern Hemisphere, trade winds blow from the northeast, while in the Southern Hemisphere, they blow from the southeast.
- **Westerlies**: Blow from the subtropical high-pressure belts towards the subpolar low-pressure belts. In the mid-latitudes, westerlies blow from the west towards the east.
- **Polar Easterlies**: Blow from the polar high-pressure belts towards the subpolar low-pressure belts. These winds are cold and dry.

3.4.3. Working of these Systems

The global pressure belts and wind systems work together to create a circulation pattern in the atmosphere. Air rises at the equator, creating a low-pressure belt, and sinks at the subtropics, creating a high-pressure belt. This circulation pattern drives the trade winds, westerlies, and polar casterlies.

3.4.4. Their Importance

Understanding global pressure belts and wind systems is crucial for predicting weather patterns, climate, and ocean currents. These patterns play a significant role in shaping regional climates, influencing precipitation, and impacting global weather phenomena like hurricanes and monsoons.

In summary, global pressure belts and wind systems are large-scale patterns of atmospheric circulation that drive the Earth's climate and weather. The pressure belts are created by uneven heating of the Earth's surface, and the wind systems are driven by pressure differences between the belts.

3.5. Daily and Seasonal Variation of Wind Speed

Daily and seasonal variations in wind speed are significant factors in weather patterns, climate, and renewable energy production. Here's a breakdown of each:

3.5.1. Daily (Diurnal) Variations

Daily variations in wind speed are primarily driven by the **heating and cooling of the Earth's surface**throughout the day.

- **Daytime:** Solar radiation heats the ground, which in turn heats the air near the surface. This warmer air rises (convection), leading to increased vertical mixing in the atmosphere. This mixing can bring down higher-speed winds from aloft, resulting in stronger surface winds during the day, typically peaking in the afternoon.

- **Nighttime:** As the sun sets, the ground cools, and the air near the surface also cools and becomes more stable. This stability reduces vertical mixing, and surface friction becomes a more dominant factor. Consequently, surface wind speeds often decrease at night. However, in some coastal regions, land breezes (winds blowing from land to sea at night) can develop. Mountain and valley breezes also exhibit diurnal reversals due to differential heating and cooling of slopes and valleys.

3.5.2. Seasonal Variations

Seasonal variations in wind speed are influenced by **larger-scale weather systems and temperature differences** that change throughout the year.

- **Mid-latitudes:** Many mid-latitude regions experience stronger winds during the late fall, winter, and spring due to more intense and frequent storm systems (like mid-latitude cyclones) driven by larger temperature gradients between the poles and the equator. Summer often sees weaker winds as these storm systems are generally less intense and less frequent.
- **Coastal Regions:** Seasonal differences in the temperature contrast between land and sea can enhance or weaken coastal breezes, affecting overall wind speeds.
- **Monsoon Regions:** Areas with monsoon climates can see significant seasonal shifts in wind speed and direction. For example, the Indian subcontinent experiences strong southwesterly winds during the summer monsoon.

3.5.3. Factors Influencing Wind Speed Variations

Several factors contribute to both daily and seasonal variations in wind speed:

- **Pressure Gradients:** The ratio of difference in atmospheric pressure between two points to the distance between the two points is called pressure gradient. Differences in air pressure cause air to move from high to low pressure. The greater the pressure gradient, the stronger the wind speed.These pressure gradients change daily and seasonally.
- **Temperature Differences:** Uneven heating of the Earth's surface (both daily and seasonally) leads to pressure differences, driving wind.
- **Coriolis Effect:** The Earth's rotation deflects moving air (to the right in the Northern Hemisphere and to the left in the Southern Hemisphere). This effect influences the direction of the wind more significantly at

larger scales but also plays a role in the overall wind patterns that vary seasonally.

- **Friction:** The Earth's surface slows down the wind; especially near the ground.This effect is more pronounced during the day when vertical mixing is weaker near the surface at night. The type of surface (e.g., smooth water vs. rough terrain) also affects friction.
- **Local Geography:** Features like mountains, valleys, and coastlines can create localized wind patterns and modify regional wind speeds both daily and seasonally.

Understanding these daily and seasonal variations is crucial for various applications, including weather forecasting, climate modeling, agriculture, and optimizing the efficiency of wind energy generation.

3.6. Cyclones

A **cyclone** is a large-scale weather system characterized by a **low-pressure center** and **winds spiraling inward**. The direction of this spiraling wind depends on the hemisphere:

- **Northern Hemisphere:** Winds rotate **counterclockwise**.
- **Southern Hemisphere:** Winds rotate **clockwise**.

Cyclones are often associated with strong winds, heavy rainfall, and severe weather conditions. Depending on their origin and intensity, they are known by different names around the world (e.g., hurricanes in the Atlantic and eastern Pacific, typhoons in the western Pacific).

Cyclones are classified into several types based on their location, intensity, and characteristics. Here are some of the main forms:

- **Tropical Cyclones**: These form over warm ocean waters in the tropics and are known as hurricanes, typhoons, or cyclones depending on the location.
- **Extratropical Cyclones**: These form outside the tropics and are associated with fronts and low-pressure systems.
- **Subtropical Cyclones**: These have characteristics of both tropical and extratropical cyclones.
- **Polar Cyclones**: These form at high latitudes and are associated with cold air masses.
- **Mesocyclones**: These are small-scale cyclones that form in association with thunderstorms and can produce tornadoes.

In terms of intensity, cyclones can be classified as

- **Tropical Depression**: Maximum sustained winds of 38 mph or less.
- **Tropical Storm**: Maximum sustained winds of 39-73 mph.
- **Cyclone/Hurricane/Typhoon**: Maximum sustained winds of 74 mph or higher.

Each type has distinct characteristics and impacts different regions.

3.6.1. Structure of Cyclone

A cyclone, also known as a hurricane or typhoon depending on the region, has a distinct structure (Fig. 3.5).

Fig. 3.5: Structure of cyclone

The main components are

- **The Eye:** This is the central, relatively calm area of the cyclone. It typically has light winds and sometimes clear skies. The eye is a region of low surface pressure and warmer temperatures aloft due to sinking air. The diameter of the eye usually ranges from 30 to 65 kilometers (20 to 40 miles).
- **The Eyewall:** Surrounding the eye is the eyewall, a ring of intense thunderstorms. This is the most violent part of the cyclone, with the strongest winds, heaviest rain, and towering clouds. Changes in the eyewall can affect the cyclone's intensity.
- **Spiral Rainbands:** These are bands of clouds and showers that spiral outward from the eyewall. They can produce heavy rainfall and sometimes contain thunderstorms and even tornadoes, particularly in the outer regions of the cyclone. The extent and organization of these rainbands can give clues about the cyclone's intensity.

In summary, a cyclone is a system of rotating air with the calm **eye** at the center, surrounded by the intense **eyewall**, and further out by **spiral rainbands**. Air spirals inward at low levels, rises in the eyewall and rainbands, and flows outward at the upper levels. In the Northern Hemisphere, this rotation is counter-clockwise, and clockwise in the Southern Hemisphere, due to the Coriolis Effect.

3.6.2. Track of a cyclones: The track of a cyclone refers to the path taken by the center of the cyclone as it moves over the Earth's surface

This path is usually depicted on a map, showing the successive positions of the cyclone's center (often the eye, if one is present) at different points in time.

Understanding the track is crucial for forecasting where the cyclone will go and which areas will be affected by its strong winds, heavy rainfall, and storm surge.

In simpler terms, it's like drawing a line on a map that shows where the "middle" of the cyclone has been and where it is predicted to go.

To locate the track of a cyclone, meteorologists use a combination of sophisticated tools and techniques:

- **Satellite Imagery:** Geostationary satellites provide continuous images in visible and infrared wavelengths, allowing for the identification of the cyclone's center and the monitoring of its movement over time. By observing cloud patterns and their evolution, the initial track can be estimated.
- **Weather Radar:** Coastal radars can detect the precipitation associated with a cyclone and track its movement as it approaches land. Doppler radar can further provide information on the wind speeds within the storm, which can help in inferring the location of the center.
- **Aircraft Reconnaissance:** Specially equipped aircraft fly into the cyclone to get direct measurements of the storm's center, including pressure, wind speed, and location. This provides highly accurate positional data.
- **Surface Observations:** Reports from land-based weather stations, ships, and buoys offer ground-level data on wind, pressure, and rainfall, which can help confirm the cyclone's location, especially as it nears land.
- **Numerical Weather Prediction Models:** Computer models ingest all available observational data to simulate the atmosphere and forecast the cyclone's future path. These models are crucial for predicting the track several days in advance.

The process involves continuously analyzing data from these sources to pinpoint the current center of the cyclone and then using the models and historical data to forecast its future path. Meteorological centers around the world, like the India Meteorological Department (IMD) for your region, integrate this information to determine and disseminate the cyclone's track.

3.6.3. Origin, Intensification and Dissipation of Cyclones

Cyclones are powerful weather systems that form over warm oceanic waters in tropical and subtropical regions. Let's break down their origin, intensification, and dissipation.

- **Origin of Cyclones:** Cyclones form when specific meteorological conditions come together, including
  - **Warm Ocean Waters**: Sea surface temperatures need to be at least 26.5°C (80°F) to fuel rapid evaporation and cloud formation.
  - **Low-Pressure System**: Air rises, forming clouds, and the Coriolis Effect creates cyclonic circulation around the low-pressure center.
  - **Moist Air and High Humidity**: Warm, moist air contributes to cloud formation and storm intensification.
  - **Favorable Wind Conditions**: Minimal vertical wind shear is essential for cyclone development.
- **Intensification of Cyclones:** Cyclones intensify when they have access to:
  - **Warm Ocean Waters**: Continued warmth fuels further evaporation and cloud formation.
  - **High Ocean Heat Content**: Allows storms to achieve higher intensities.
  - **Symmetric Outflow**: Strong outflow leads to faster intensification.
  - **Low Vertical Wind Shear**: Minimal shear enables the storm to maintain its structure.
- **Dissipation of Cyclones:** Cyclones dissipate when they
  - **Make Landfall**: Loss of moisture supply, increased surface friction, and dry air intrusion weaken the storm.
  - **Encounter Dry Air**: Dry air disrupts cloud formation and weakens the storm system.
  - **Interact with Mountains**: Disrupted circulation and orographic lifting can fragment the storm's structure.
  - **Move over Cold Waters**: Cyclones need warm waters to sustain themselves; moving over cold waters cuts off their energy source.

Normally, the cyclone intensifies as long as it remains in ocean, as the water surface offers less friction and is favorable for strong circulation of winds.

When the cyclones cross sea coast, they get weakened due to friction offered by the land surface and gradually dissipate when it moves over the land.

3.7. Depressions

In meteorology, a **depression** (also known as a low-pressure area or simply a "low") is a region where the atmospheric pressure is lower than that of the surrounding locations. Air tends to flow towards these low-pressure areas.

- Depressions are associated with rising air, which can lead to cloud formation and precipitation.
- They can range in intensity from weak, causing just cloudy and unsettled weather, to strong, bringing significant wind and rain.

3.7.1. Shape of Depressions in Isobaric

On isobaric charts (maps showing lines of equal atmospheric pressure called isobars), depressions are typically represented by a series of **roughly circular or oval-shaped closed isobars**, with the lowest pressure at the center. The isobars are closer together when the pressure gradient is steep, indicating stronger winds.

3.7.2. Differentiate Between Cyclones and Depressions

The terms "cyclone" and "depression" are related but often denote different intensities of a low-pressure system, especially in the context of tropical weather. Here's a general way to differentiate them:

- **Intensity (Wind Speed)**
    - A **depression** generally has lower wind speeds. Different meteorological agencies have specific thresholds, but typically, a tropical depression has maximum sustained winds below a certain limit (e.g., below 39 mph or 63 km/h).
    - A **cyclone** (which includes tropical storms, hurricanes, typhoons, etc., depending on the region and wind speed) has stronger, gale-force winds exceeding these thresholds.
- **Organization**
    - A **depression** might have a less organized structure of clouds and thunderstorms.
    - A **cyclone** typically exhibits a more organized circulation, often with a well-defined center and spiral rainbands. More intense cyclones can develop an eye.

- **Nomenclature**
  - Once a tropical depression intensifies beyond a certain wind speed, it is usually upgraded to a tropical storm (or a similar classification depending on the basin) and given a name, signifying a greater intensity than a depression. The common rule is that the name list is proposed by the National Meteorological and Hydrological Services (NMHSs) of WMO Members of a specific region, and approved by the respective tropical cyclone regional bodies at their annual/ biennual sessions.

In essence, a cyclone is a more intense and organized form of a low-pressure system compared to a depression. A depression can be a precursor to a cyclone if conditions favor intensification.

In some general meteorological contexts outside of tropical systems, "depression" might simply refer to any area of low pressure. However, when discussing tropical weather, the distinction based on intensity is usually made.

3.8. Anticyclones

An **anticyclone** is a large-scale circulation of winds around a central region of **high atmospheric pressure**, opposite to that of a cyclone (Fig. 3.6). It can be further explained as:

- **High Pressure:** At the center of an anticyclone, the atmospheric pressure is higher than in the surrounding areas.
- **Sinking Air:** In an anticyclone, air from higher up in the atmosphere descends towards the surface. This sinking air generally leads to stable atmospheric conditions.

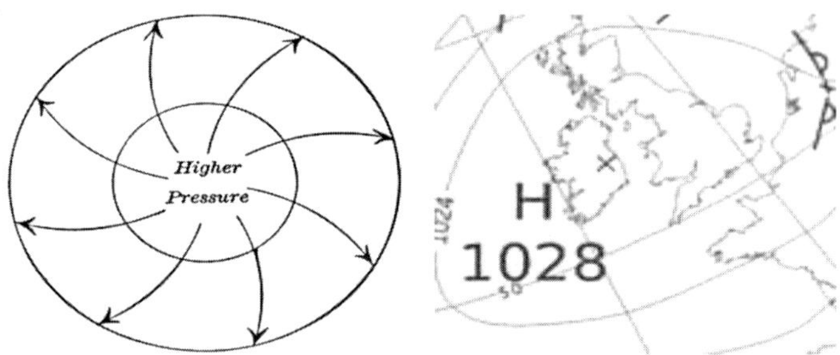

Fig. 3.6: Anticyclone

- **Wind Circulation:** Due to the Coriolis effect:
  - In the **Northern Hemisphere**, winds around an anticyclone flow **clockwise** and outward from the high-pressure center.

- In the **Southern Hemisphere**, winds flow **counter-clockwise** and outward.

3.8.1. Characteristics and Weather Associated with Anticyclones

- **Fair Weather:** Generally, anticyclones are associated with settled, fine weather, often bringing clear skies and calm conditions because the sinking air inhibits cloud formation.
- **Light Winds:** The pressure gradient in an anticyclone is usually weak, leading to gentle winds. Isobars (lines of equal pressure) on a weather map are typically widely spaced in an anticyclone.
- **Temperature Variations**
    - **Summer:** Can bring hot, sunny, and dry weather, sometimes leading to heatwaves.
    - **Winter:** Often results in cold, clear nights (due to radiative cooling) and can lead to frost and fog, especially in calm conditions. Sometimes, persistent high pressure in winter can cause prolonged periods of cold weather and low-lying cloud or "anticyclonic gloom."

In isobaric charts, anticyclones are represented by closed, roughly circular or oval isobars, with the highest pressure at the center.

In essence, while cyclones are "lows" bringing unsettled weather, anticyclones are "highs" typically associated with more stable and often pleasant conditions.

3.9. Land and Sea Breezes

- **Formation of Land and Sea Breezes:** Land and sea breezes are local winds that occur near coastlines due to the **differential heating and cooling rates of land and water** (Fig. 3.7).

3.9.1. Sea Breeze (occurs during the day)

- During the day, the **land heats up much faster** than the sea because land has a lower specific heat capacity.
- The air above the land gets heated, becomes less dense, and rises, creating a **low-pressure area** over the land.
- The air over the sea, which heats up more slowly, remains cooler and denser, resulting in a **high-pressure area** over the sea.
- Air flows from the **high pressure over the sea towards the low pressure over the land**, creating a **sea breeze**. This cool breeze from the sea moves inland.

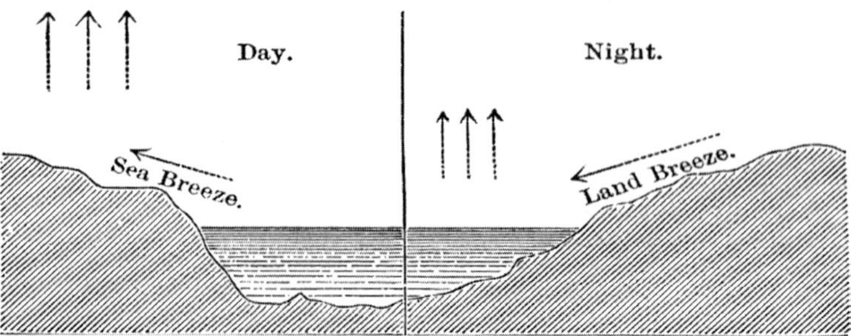

Fig. 3.7: Land and sea breeze

3.9.2. Land Breeze (occurs during the night)

- At night, the **land cools down much faster** than the sea, which retains heat longer.
- The air above the land becomes cooler and denser, resulting in a **high-pressure area** over the land.
- The air over the sea remains warmer and less dense, creating a **low-pressure area** over the sea.
- Air flows from the **high pressure over the land towards the low pressure over the sea**, creating a **land breeze**. This gentle breeze blows from the land towards the sea.

3.9.3. Effects of Land and Sea Breezes

Land and sea breezes have several effects, primarily on coastal areas:

- **Temperature Moderation:** They help to moderate the temperature in coastal areas. Sea breezes bring cooler air inland during hot days, while land breezes bring slightly warmer air from the land out to sea during cooler nights. This reduces the extreme temperature variations experienced inland.
- **Humidity:** Sea breezes are usually more humid as they bring air from over the water. Land breezes tend to be drier.
- **Local Weather Patterns**
  - Sea breeze fronts (the boundary between the advancing sea breeze and the warmer inland air) can sometimes trigger the development of clouds and even thunderstorms as the cooler, denser marine air undercuts the warmer land air, causing it to rise.
  - Land breezes can sometimes cause the development of clouds or showers offshore if there is enough moisture and instability.

- **Pollution Dispersion:** These local winds can help to disperse pollutants in coastal areas.
- **Aviation and Sailing:** Sea breezes can be stronger than land breezes and are important for activities like gliding and sailing.

In summary, land and sea breezes are important local wind systems driven by the different heating and cooling properties of land and water, significantly influencing the weather and climate of coastal regions.

4

Solar Radiation

The process through which heat energy is transferred from a body at higher temperature to a body at lower temperature without any material medium between the two bodies is x called radiation. The outer surface of the the Sun is at higher temperature compared to the surface of the earth. Solar radiation is the electromagnetic energy emitted by the Sun. It is the primary source of energy for the Earth's climate system, driving atmospheric and oceanic circulation, influencing weather patterns, and sustaining life through photosynthesis.2 This energy travels through space as electromagnetic waves, encompassing a broad spectrum of wavelengths, from high-energy gamma rays and X-rays to ultraviolet, visible, infrared, and radio waves. The portion of this spectrum that reaches the Earth's surface is crucial for a multitude of processes, making the study of solar radiation fundamental to understanding our planet's environment. This introduction will lay the groundwork for exploring the characteristics, distribution, and interaction of solar radiation with the Earth's atmosphere and surface.

4.1. Wave Length of Radiation

The wavelength of radiation refers to the distance between two consecutive peaks or troughs of a wave. It's a measure of the length of one complete cycle of the wave (Fig. 4.1).

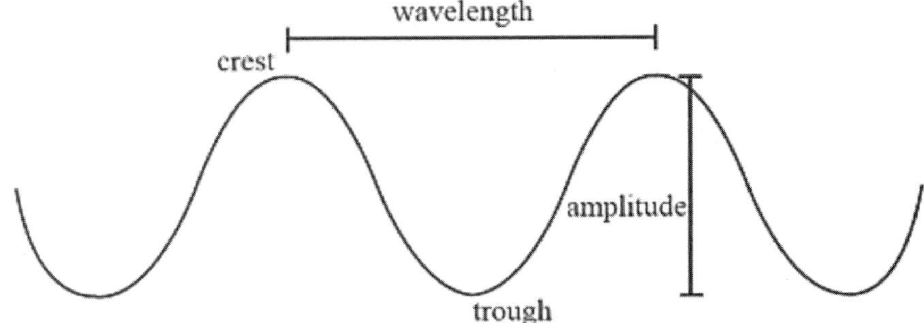

Fig. 4.1: Wave length of radiation

Imagine a wave in the ocean. The wavelength would be the distance from one crest (peak) to the next crest. In the context of radiation, such as light or other forms of electromagnetic radiation, wavelength determines the type and characteristics of the radiation. The wavelength of radiation is often represented by the Greek letter Lambda (λ).

So, λ (lambda) = wavelength

In equations, you'll often see λ used to denote the wavelength of radiation, such as in the formula:

$$c = \lambda \nu$$

where c is the speed of light, λ is the wavelength, and ν is the frequency of radiation. In general, the intensity of radiation decreases with increase in wave length. The short wave radiation will be of higher intensity compared to long wave radiation.

4.2. Electromagnetic Spectrum

The **electromagnetic spectrum** is the entire range of all types of electromagnetic (EM) radiation. Radiation is energy that travels and spreads out as it goes – the visible light from a lamp and the radio waves1 from a radio station are both types of electromagnetic radiation. The electromagnetic spectrum encompasses all forms of EM radiation, ordered by frequency and wavelength (Fig. 4.2).

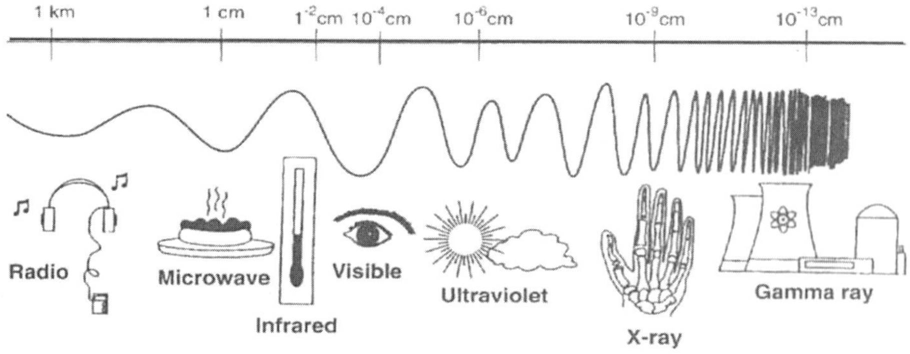

Fig. 4.2: Electromagnetic spectrum

The main types of electromagnetic radiation, from longest wavelength (lowest frequency and energy) to shortest wavelength (highest frequency and energy), are:

- **Radio Waves:** These have the longest wavelengths and are used for communication (radio, TV, cell phones), radar, and astronomy.

- **Microwaves:** Shorter wavelengths than radio waves, used in microwave ovens, satellite communications, and some radar systems.
- **Infrared (IR):** Often associated with heat. Used in thermal imaging, remote controls, and some types of astronomy.
- **Visible Light:** The narrow range of wavelengths that the human eye can detect. It includes all the colors of the rainbow.
- **Ultraviolet (UV):** Shorter wavelengths than visible light. Can cause tanning and sunburn. Most UV radiation from the sun is absorbed by the Earth's atmosphere.
- **X-rays:** Even shorter wavelengths, used in medical imaging and security scanning because they can penetrate soft tissues.
- **Gamma Rays:** The shortest wavelengths and highest energy. Produced in nuclear reactions and some astronomical events. Used in cancer treatment and industrial applications.

4.3. Solar Spectrum

The **solar spectrum** is the range of electromagnetic radiation emitted by the Sun. This radiation spans a broad range of wavelengths, from very short gamma rays and X-rays to ultraviolet, visible, infrared, and radio waves (Fig. 4.3). However, the majority of the Sun's energy output falls within the ultraviolet, visible, and infrared portions of the spectrum.

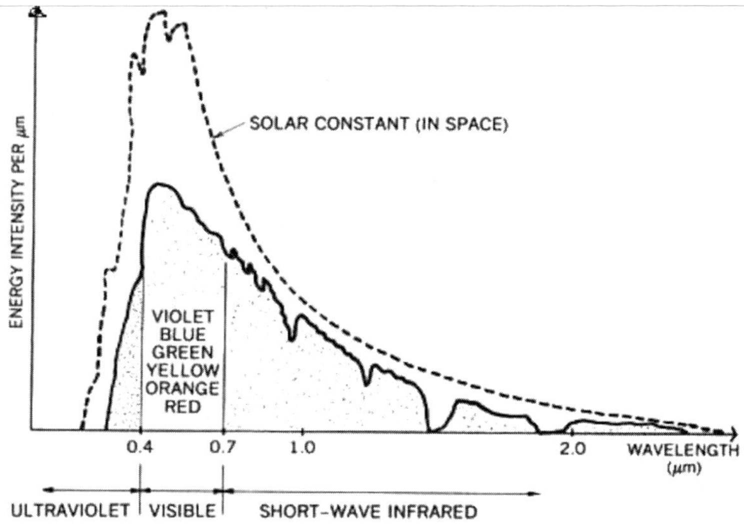

Fig. 4.3: Solar spectrum at the Earth's surface

Here are some key aspects of the solar spectrum

- **Distribution of Energy:** The Sun emits energy across all these wavelengths, but the intensity varies. The solar spectrum is often depicted as a graph showing the intensity of radiation at each wavelength. The peak of this spectrum falls within the visible light range, which is why our eyes are most sensitive to these wavelengths.
- **Components**
    - **Ultraviolet (UV):** Shorter wavelengths than visible light. The Sun emits UV-A, UV-B, and UV-C radiation, but the Earth's atmosphere (especially the ozone layer) absorbs most of the harmful UV-B and all of UV-C.
    - **Visible:** The portion of the spectrum (approximately 400-700 nanometers) that humans can see as colors (violet, indigo, blue, green, yellow, orange, red).
    - **Infrared (IR):** Longer wavelengths than visible light, often felt as heat.
- **Solar Irradiance:** This refers to the power of solar radiation per unit area. The **solar constant** is the total solar irradiance at the top of Earth's atmosphere. However, the spectrum of this radiation shows how this total energy is distributed across different wavelengths.
- **Atmospheric Effects:** As solar radiation passes through the Earth's atmosphere, certain wavelengths are absorbed or scattered by gases (like ozone, water vapor, carbon dioxide) and particles.This is why the solar spectrum that reaches the Earth's surface is different from the spectrum at the top of the atmosphere. For example, the atmosphere strongly absorbs certain UV wavelengths.
- **Absorption Lines:** The solar spectrum also contains absorption lines (Fraunhofer lines), which are dark lines at specific wavelengths. These are caused by the absorption of light by elements in the Sun's atmosphere. Analyzing these lines helps scientists determine the composition of the Sun.
- In essence, the solar spectrum is a fingerprint of the Sun's energy output across different wavelengths, and understanding it is crucial for studying Earth's climate, atmospheric processes, and even the Sun itself.

4.4. Nature and Properties of Solar Radiation

Solar radiation is the electromagnetic energy emitted by the Sun. It's the fundamental source of energy for Earth's climate and life. Let's explore its nature and properties:

4.4.1. Nature of Solar Radiation

- **Electromagnetic Waves:** Solar energy travels through space as electromagnetic waves. These waves have both electrical and magnetic components and can travel through the vacuum of space.
- **Broad Spectrum:** This radiation isn't just visible light; it encompasses a wide range of wavelengths, forming the electromagnetic spectrum. The Sun emits energy across this spectrum, including gamma rays, X-rays, ultraviolet (UV), visible, infrared (IR), and radio waves.
- **Energy Source:** The Sun's energy originates from nuclear fusion reactions in its core, where hydrogen atoms are converted into helium, releasing vast amounts of energy. This energy is then radiated outwards.

4.4.2. Properties of Solar Radiation

- **Wavelength:** Different types of solar radiation have different wavelengths, measured in nanometers (nm) or micrometers ($\(\mu m\)$). Shorter wavelengths (like UV and X-rays) generally carry higher energy than longer wavelengths (like infrared and radio waves). The solar spectrum that reaches Earth is primarily composed of UV, visible, and IR radiation.
- **Intensity (Irradiance):** The power of solar radiation per unit area is called irradiance, measured in Watts per square meter.
- **Transmission, Absorption, and Reflection:** When solar radiation reaches the Earth's atmosphere and surface, it can be:
  - **Transmitted:** Passing through without being altered.
  - **Absorbed:** Taken in by the atmosphere, land, or oceans, increasing their internal energy. Different gases absorb specific wavelengths (e.g., ozone absorbs UV).
  - **Reflected:** Bounced back into space (e.g., by clouds, ice, and bright surfaces). The fraction reflected is the albedo.
- **Direct vs. Diffuse Radiation**
  - **Direct radiation** comes straight from the Sun.
  - **Diffuse radiation** is sunlight that has been scattered by the atmosphere (e.g., by clouds, dust, and air molecules).
- **Variability:** The amount of solar radiation received at a particular location varies with:
  - **Latitude:** Equator receives more direct sunlight.
  - **Time of day:** Highest at noon.

- **Season:** Due to Earth's tilt.
- **Atmospheric conditions:** Clouds, pollution, etc.

In summary, solar radiation is a fundamental electromagnetic energy source with a spectrum of wavelengths, varying intensity, and interacts with the Earth's system through transmission, absorption, and reflection, driving many of our planet's processes.

4.5. Classification of Waves and Types

The electromagnetic radiation is classified into different types of radiation depending upon the wavelength of radiation.

- X-rays: 0.01 nm to 10 nm
- Ultraviolet (UV) radiation: 10 nm to 400 nm
- Visible light: 400 nm (violet) to 700 nm (red)
- Infrared (IR) radiation: 700 nm to 1 millimeter (mm)
- Microwaves: 1 mm to 1 meter (m)
- Radio waves: 1 m to thousands of kilometers

Here are some common conversion factors for units of wavelength

- 1 nanometer (nm) = 10^-9 meters (m)
- 1 micrometer (μm) = 10^-6 meters (m) = 1000 nm
- 1 angstrom (Å) = 0.1 nm = 10^-10 m
- 1 millimeter (mm) = 10^-3 meters (m)

Some common conversions:

- nm to μm: divide by 1000 (e.g., 500 nm = 0.5 μm)
- μm to nm: multiply by 1000 (e.g., 2 μm = 2000 nm)
- nm to Å: multiply by 10 (e.g., 5 nm = 50 Å)
- Å to nm: divide by 10 (e.g., 50 Å = 5m)

4.6. Solar Constant

The solar constant is approximately 1353 watts per square meter (W/m^2). It's the amount of solar energy received per unit area at the top of Earth's atmosphere when the planet is at its average distance from the Sun. This value can vary slightly due to changes in Earth's orbit and solar activity.

4.7. Short wave, Long wave and Thermal Radiation

- **Shortwave radiation**: Refers to the visible and ultraviolet (UV) radiation emitted by the Sun, with wavelengths roughly between 0.1-4 micrometers. This radiation is absorbed by the Earth's surface and atmosphere.

- **Longwave radiation**: Refers to the infrared (IR) radiation emitted by the Earth's surface and atmosphere, with wavelengths roughly between 4-100 micrometers. This radiation is a result of the Earth's surface and atmosphere heating up after absorbing shortwave radiation.
- **Thermal radiation**: A broader term that encompasses all forms of electromagnetic radiation emitted by objects due to their temperature. Both shortwave and longwave radiation can be forms of thermal radiation, but in the context of Earth's energy balance, thermal radiation often specifically refers to the longwave IR radiation emitted by the planet.

It may be further explained as

4.7.1. Shortwave Radiation

- Comes from the Sun
- Wavelengths: 0.1-4 micrometers (UV, visible light, and near-infrared)
- Includes
  - Ultraviolet (UV) radiation (0.1-0.4 micrometers)
  - Visible light (0.4-0.7 micrometers)
  - Near-infrared radiation (0.7-4 micrometers)
- Absorbed by Earth's surface, oceans, and atmosphere

4.7.2. Longwave Radiation

- Emitted by the Earth's surface and atmosphere
- Wavelengths: 4-100 micrometers (infrared radiation)l
- Result of Earth's surface and atmosphere heating up after absorbing shortwave radiation
- Trapped by greenhouse gases, contributing to the greenhouse effect

4.7.3. Thermal Radiation

- Emitted by all objects due to their temperature
- Encompasses a wide range of wavelengths, including shortwave and longwave radiation
- Earth's surface and atmosphere emit thermal radiation in the form of longwave IR radiation
- Plays a crucial role in Earth's energy balance and climate

In the context of Earth's energy balance, the interaction between shortwave and longwave radiation is crucial. The balance between the two determines the planet's climate and temperature.

4.8. Radiation Balance and Depletion of Solar Radiation

The radiation balance refers to the equilibrium between the amount of solar radiation absorbed by the Earth and the amount of radiation emitted back into space. This balance determines the Earth's climate and temperature.

4.8.1. Components of Radiation Balance

- **Incoming Solar Radiation**: Shortwave radiation from the Sun.
- **Absorbed Radiation**: Radiation absorbed by the Earth's surface and atmosphere.
- **Outgoing Longwave Radiation**: Longwave radiation emitted by the Earth's surface and atmosphere.

4.8.2. Depletion of Solar Radiation

Depletion of solar radiation occurs when the amount of solar radiation reaching the Earth's surface is reduced as it travels through the atmosphere. This can happen due to:

- **Atmospheric Scattering**: Aerosols, pollutants, and particles in the atmosphere scatter solar radiation, reducing its intensity.
- **Cloud Cover**: Clouds can block or reflect solar radiation, reducing the amount that reaches the surface.
- **Aerosol Optical Depth**: High levels of aerosols in the atmosphere can reduce solar radiation.
- **Earth's Orbital Variations**: Changes in Earth's orbit can affect the amount of solar radiation received.

4.8.3. Consequences

- **Climate Change**: Changes in radiation balance can impact global temperatures and climate patterns.
- **Reduced Solar Energy**: Depletion of solar radiation can affect solar power generation and ecosystems that rely on sunlight.

4.8.4. Factors Influencing Radiation Balance

- **Greenhouse Gases**: Trap longwave radiation, contributing to global warming.
- **Aerosols**: Can cool or warm the planet, depending on their type and location.
- **Land Use Changes**: Can alter the amount of solar radiation absorbed by the surface.

Understanding radiation balance and depletion of solar radiation is crucial for predicting climate patterns, managing solar energy resources, and mitigating the effects of climate change.

4.9. Greenhouse Gases

Greenhouse gases (GHGs) are gases in the Earth's atmosphere that absorb and emit infrared radiation, trapping heat and contributing to the greenhouse effect (Fig. 4.4).

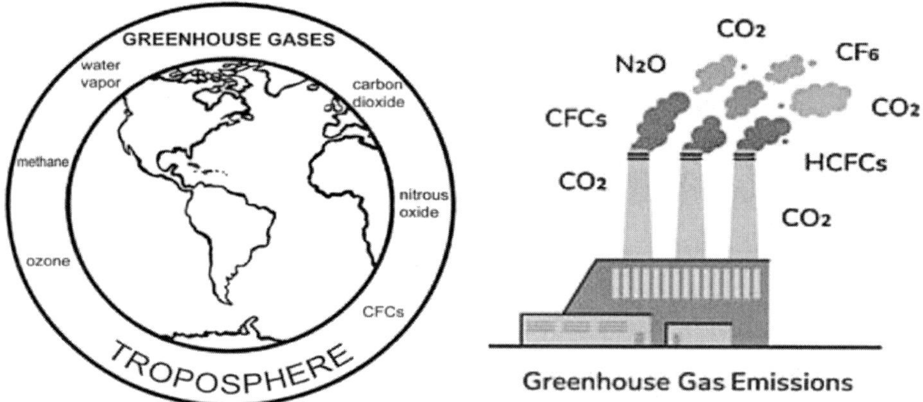

Fig. 4.4: Greenhouse gases

The main GHGs are

- **Carbondioxide (CO_2)**: Released through fossil fuel combustion, deforestation, and land-use changes.
- **Methane (CH_4)**: Released through agriculture (especially rice and cattle farming), natural gas production and transport, and landfills.
- **Nitrous oxide (N_2O)**: Released through agriculture, industrial processes, and burning fossil fuels.
- **Fluorinated gases (F-gases)**: Released through air conditioning, refrigeration, and electrical equipment.
- **Ozone (O_3)**: Forms in the stratosphere through the interaction of ultraviolet radiation and oxygen molecules.
- **Water vapor (H_2O)**: The most abundant GHG, playing a crucial role in the Earth's climate system.

These gases trap heat in the atmosphere, leading to global warming and associated climate change impacts, such as:

- Rising global temperatures
- More extreme weather events
- Sea-level rise
- Changes in precipitation patterns

Human activities, especially burning fossil fuels and deforestation, have significantly increased GHG concentrations, accelerating global warming.

4.10. Aerosols

Aerosols are tiny particles or droplets suspended in the air. They can be solid or liquid and come from various sources, including:

4.10.1. Natural Sources

- Dust
- Sea salt
- Volcanic ash
- Plant pollen

4.10.2. Human Activities

- Fossil fuel combustion (e.g., soot, particulate matter)
- Industrial processes
- Vehicle emissions
- Agricultural activities

4.10.3. Aerosols can affect

- **Climate**: Aerosols can reflect sunlight, cooling the planet (e.g., sulfates) or absorb sunlight, warming the planet (e.g., black carbon).
- **Air Quality**: Aerosols can negatively impact human health, especially respiratory health.
- **Weather Patterns**: Aerosols can influence cloud formation and precipitation.

4.10.4. Examples of Aerosols Include

- **Sulfate aerosols**: From fossil fuel combustion and volcanic activity
- **Black carbon**: From incomplete combustion of fossil fuels and biomass
- **Dust**: From deserts, construction, and agricultural activities
- **Smoke**: From wildfires and biomass burning
- Aerosols play a complex role in the Earth's climate system, and their impacts can vary depending on their type, size, and location.

4.11. Net Radiation

Net radiation is the balance between the incoming and outgoing radiation at the Earth's surface or top of the atmosphere. It represents the amount of energy available for various processes, such as heating the surface, evaporating water, or driving atmospheric circulation.

4.11.1. Equation for Net Radiation

The net radiation (Rn) equation is:

Rn = (Incoming Shortwave Radiation) - (Reflected Shortwave Radiation) + (Incoming Longwave Radiation) - (Outgoing Longwave Radiation)

Mathematically, this can be represented as:

$Rn = (S\downarrow - S\uparrow) + (L\downarrow - L\uparrow)$

Where:

- Rn = Net radiation
- $S\downarrow$= Incoming shortwave radiation (solar radiation)
- $S\uparrow$= Reflected shortwave radiation (albedo)
- $L\downarrow$= Incoming longwave radiation (atmospheric radiation)
- $L\uparrow$= Outgoing longwave radiation (surface-emitted radiation)

This equation helps understand the energy balance at the Earth's surface or top of the atmosphere, which is crucial for climate modeling, weather forecasting, and understanding various Earth system processes.

4.11.2. Albedo

Albedo is a measure of the amount of solar radiation that is reflected by a surface, compared to the amount of radiation that hits it. It's a dimensionless value between 0 and 1, where:

- 0 represents a perfect absorber (no reflection)
- 1 represents a perfect reflector (total reflection)

4.11.3. Albedo Values for Different Surfaces

Here are some typical albedo values for various surfaces:

- **Low Albedo (Dark Surfaces)**
  - Water (deep ocean): 0.05-0.10
  - Asphalt: 0.05-0.10
  - Soil: 0.05-0.20
  - Forests: 0.10-0.20

- **Moderate Albedo (Medium Surfaces)**
    - Grasslands: 0.20-0.30
    - Crops: 0.20-0.30
    - Urban areas: 0.15-0.30
- **High Albedo (Light Surfaces)**
    - Fresh snow: 0.80-0.90
    - Clouds: 0.60-0.90
    - Sand: 0.40-0.60
    - Ice: 0.50-0.70

4.11.4. Importance of Albedo

Albedo plays a crucial role in determining the Earth's energy balance and climate. Changes in albedo can impact:

- **Global Temperature**: Changes in albedo can influence the amount of solar radiation absorbed, affecting global temperatures.
- **Climate Feedbacks**: Albedo changes can feedback on climate, influencing cloud formation, ice sheet extent, and more.
- **Weather Patterns**: Albedo can impact local weather patterns, such as temperature and precipitation.

Understanding albedo is essential for climate modeling, weather forecasting, and land surface characterization.

5

Light - A Comprehensive Overview

Light is a form of electromagnetic radiation that is visible to the human eye. It is a crucial aspect of our daily lives, and its importance extends beyond just visibility (Fig. 5.1). Light plays a vital role in various fields, including agriculture, energy, health, and more.

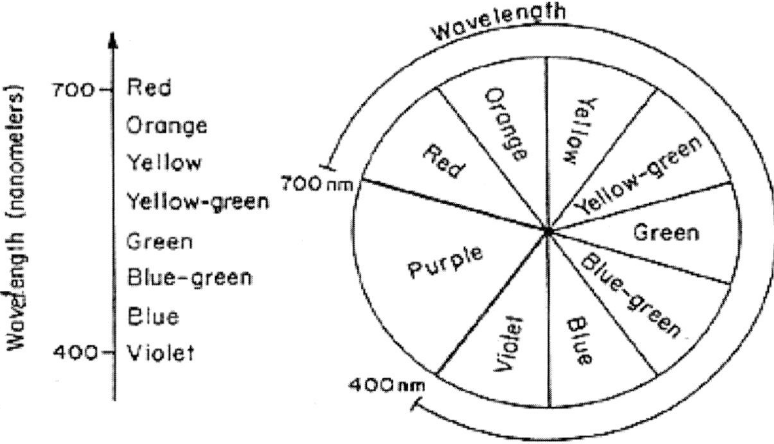

Fig. 5.1: Science of light

5.1. Properties of Light

- **Wavelength**: Light consists of different wavelengths, which determine its color. The visible spectrum of light includes colors ranging from approximately 380 nanometers (violet) to 780 nanometers (red).
- **Intensity**: Light intensity affects its brightness and energy. Higher intensity light is more energetic and can have a greater impact on various processes.
- **Direction**: Light travels in straight lines and can be directed or focused. This property is essential for applications like lasers, optics, and solar concentrators.

5.2. Types of Light

- **Natural Light**: Natural light comes from the sun, moon, and stars. Sunlight is the primary source of natural light and is essential for many processes, including photosynthesis and vitamin D production in humans.
- **Artificial Light**: Artificial light is generated by human-made sources, such as lamps, light-emitting diodes (LEDs), and lasers. These sources can mimic natural light or provide specific wavelengths and intensities for various applications.

5.3. Importance of Light in Different Fields

- **Agriculture**: Light is essential for plant growth and development. Photosynthesis, the process by which plants convert CO_2 and water into glucose and oxygen, relies on light energy. Different wavelengths and intensities of light can impact plant growth, flowering, and fruiting.
- **Energy**: Light is a source of renewable energy through solar power. Photovoltaic cells convert sunlight into electricity, providing a clean and sustainable energy source.
- **Health**: Light therapy is used to treat conditions like seasonal affective disorder (SAD), which is characterized by depression and lethargy during periods of low sunlight. Light exposure can also regulate circadian rhythms and improve mood.
- **Technology**: Light is used in various technologies, including lasers, optical communication systems, and medical imaging.

5.4. Light Spectrum and Its Applications

The light spectrum consists of different wavelengths, each with unique properties and applications:

- **Ultraviolet (UV) Light**: UV light has a shorter wavelength than visible light and is used in applications like disinfection, curing of inks and adhesives, and medical treatments.
- **Visible Light**: Visible light is essential for human vision and is used in various applications, including lighting, displays, and optical communication systems.
- **Infrared (IR) Light**: IR light has a longer wavelength than visible light and is used in applications like heating, thermal imaging, and remote sensing.

In conclusion, light is a vital aspect of our lives, and its importance extends beyond just visibility. Its properties, types, and applications make it a crucial component in various fields, including agriculture, energy, health, and technology. Understanding light and its characteristics can help us harness its potential and develop innovative solutions for various challenges.

5.5. Quality of Light

The quality of sunlight refers to its spectral composition, intensity, and duration. Sunlight is composed of various wavelengths, including:

- **Ultraviolet (UV)**: UV radiation has a shorter wavelength and can cause damage to living tissues.
- **Visible Light**: Visible light is essential for photosynthesis and human vision.
- **Infrared (IR)**: IR radiation has a longer wavelength and contributes to heat.

5.5.1. Factors Affecting Sunlight Quality

- **Time of Day**: Sunlight quality changes throughout the day, with morning and afternoon sun having a more oblique angle.
- **Season**: Sunlight quality varies with seasons, with more direct sunlight during summer months.
- **Cloud Cover**: Clouds can scatter and diffuse sunlight, reducing its intensity.
- **Atmospheric Conditions**: Air pollution, dust, and water vapor can affect sunlight quality.

5.5.2. Importance of Sunlight Quality

- **Plant Growth**: Different wavelengths and intensities of sunlight impact plant growth, flowering, and fruiting.
- **Human Health**: Sunlight exposure is essential for vitamin D production, but excessive UV radiation can cause skin damage.
- **Energy Generation**: Sunlight quality affects solar panel efficiency and energy output.

5.5.3. Applications

- **Agriculture**: Understanding sunlight quality helps optimize crop growth and development.
- **Solar Energy**: Knowledge of sunlight quality informs solar panel design and placement.
- **Health**: Awareness of sunlight quality helps individuals balance sun exposure for vitamin D production while minimizing skin damage risks.

Sunlight quality plays a vital role in various aspects of life, and understanding its characteristics can help us harness its benefits while mitigating potential risks.

5.6. Units for Measuring Light Intensity

Light intensity can be measured using various units, depending on the context and application. Here are some common units:

- **Lux (lx)**: Lux is the SI unit for measuring illuminance, which is the amount of light falling on a surface. It's commonly used to measure light intensity in everyday applications, such as lighting design and photography.
- **Watts per square meter (W/m²)**: This unit measures the power density of light, which is useful for applications like solar energy and plant growth.
- **Photosynthetic Active Radiation (PAR)**: PAR measures the light intensity in the 400-700 nanometer range, which is essential for photosynthesis.
- **Foot-candles (fc)**: Foot-candles are an older unit for measuring illuminance, still used in some applications, especially in the United States.
- **Candelas per square meter (cd/m²)**: This unit measures luminance, which is the amount of light emitted or reflected by a surface.

5.6.1. Conversion between Units

- 1 lux = 1 lumen per square meter (lm/m^2)
- 1 $W/m^2 \approx$ 119.2 lux (for sunlight)
- 1 foot-candle $\approx$ 10.764 lux

5.6.2. Applications

- Lux and foot-candles are commonly used in lighting design, architecture, and photography.
- Watts per square meter are used in solar energy, plant growth, and climate science.
- PAR is used in agriculture, horticulture, and plant physiology.

Understanding the different units for measuring light intensity can help you choose the right unit for your specific application or context.

5.7. Duration of Light

The duration of light refers to the length of time that light is present or available. In the context of natural light, it typically refers to the number of hours or minutes of daylight or sunshine.

5.7.1. Examples

- **Day Length**: The duration of daylight, which varies throughout the year and depends on the latitude and season?
- **Sunshine Hours**: The number of hours of direct sunlight, which can be affected by cloud cover, fog, or other weather conditions.

5.7.2. Importance

- **Plant Growth**: Duration of light affects plant growth, flowering, and fruiting.
- **Circadian Rhythms**: Duration of light exposure influences human and animal circadian rhythms.
- **Energy Generation**: Duration of sunlight impacts solar energy generation.

5.7.3. Measurement

Duration of light can be measured in:

- **Hours**: Number of hours of daylight or sunshine.
- **Minutes**: Number of minutes of light exposure.
- **Day Length**: Measured in hours and minutes, often used to describe seasonal changes.

Understanding the duration of light is essential in various fields, including agriculture, biology, and energy. The duration of sunlight varies significantly around the world, depending on factors like latitude, climate patterns, topography and weather systems. Accordingly sunlight duration may further be distributed as at:

- **Equator**: Receives relatively consistent sunlight throughout the year, with approximately 4,422 hours of daylight per year.
- **Tropics**: Subtropical latitudes (around 25° to 35° north/south) have the highest sunshine values, with some areas receiving over 4,000 hours of sunshine per year.
- **Desert Regions**: The Sahara Desert, Middle East and Southwestern United States are examples of regions with high sunshine duration, often exceeding 3,000 hours per year.

- **Polar Regions**: The North Pole receives up to 4,647 hours of daylight, while the South Pole receives around 4,530 hours.

Some of the sunniest places in the world include:

- **Yuma, Arizona**: Over 4,000 hours of sunshine per year (around 91% of daylight time).
- **Marsa Alam, Egypt**: Approximately 3,958 hours of sunshine per year.
- **Sahara Desert**: Receives over 4,000 hours of sunshine per year.
- Conversely, areas with lower annual sunshine hours include ;
- **Northern Europe**: Less than 1,000 hours of sunshine annually.
- **Pacific Northwest, USA**: Around 1,600 to 2,000 hours per year.
- **United Kingdom**: Approximately 1,400 hours per year.

The duration of sunlight also changes with the seasons, with the Northern Hemisphere experiencing longer days during summer solstice (June 21st) and shorter days during winter solstice (December 21st). The opposite occurs in the Southern Hemisphere.

Effect of light on crop production may further be elaborated. Light isn't just about providing energy for plants; it's a crucial environmental cue that governs a wide array of processes from the molecular level up to the overall architecture and productivity of a crop (Fig. 5.2).

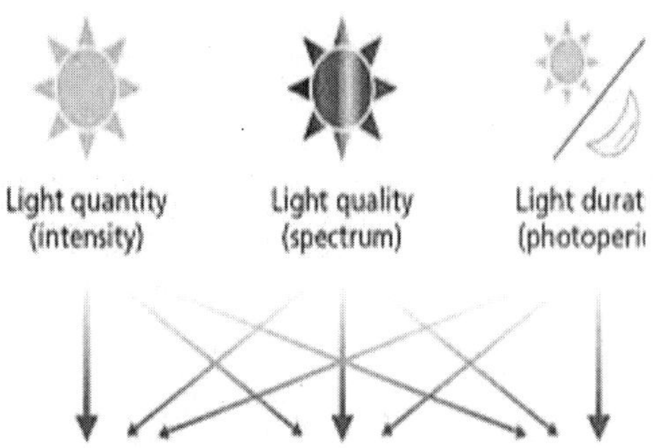

Fig. 5.2: Effect of light on crop production

5.7.4. Key Aspects with More Detail are

- **Photosynthesis: Capturing Light Energy** - At its core, the most fundamental impact of light on crop production is through **photosynthesis**. This is the biochemical process where plants, algae, and some bacteria convert light energy into chemical energy in the form of glucose (a sugar).
    - **Light Intensity:** Think of light intensity as the "brightness" of the light.

- **Low Light:** When light intensity is low, the rate of photosynthesis is directly proportional to the amount of light available. Imagine a solar panel in dim light – it won't generate much electricity. Similarly, a plant in shade will have a lower rate of sugar production, leading to slower growth, weaker stems, and often larger, thinner leaves adapted to capture more of the limited light.
- **Optimal Light:** As light intensity increases, the rate of photosynthesis increases until it reaches a plateau known as the **light saturation point**. At this point, the plant's photosynthetic machinery is working at its maximum capacity, and more light won't necessarily increase the rate. Different plant species have different light saturation points.
- **High Light:** Beyond the saturation point, excessively high light intensity can actually be detrimental. It can lead to **photoinhibition**, where the photosynthetic apparatus is damaged by too much energy. This can result in reduced photosynthetic efficiency and even leaf damage (sunscald). Plants adapted to high light environments often have mechanisms to dissipate excess energy.

- **Light Quality (Spectrum):** Light is composed of different wavelengths, which we perceive as different colors. Plants have pigments, primarily **chlorophylls** (a and b) and **carotenoids**, that absorb light most effectively in specific regions of the spectrum:
  - **Red Light (around 600-700 nm):** Chlorophylls strongly absorb red light, making it highly effective for photosynthesis.
  - **Blue Light (around 400-500 nm):** Chlorophylls also absorb blue light well, contributing significantly to photosynthesis. Blue light also plays a crucial role in other developmental processes like stomatal opening and stem elongation.
  - **Green Light (around 500-600 nm):** Plants reflect most green light, which is why they appear green. While traditionally thought to be less effective, some studies show that green light can penetrate deeper into the leaf canopy and contribute to photosynthesis in lower leaves.
  - **Far-Red Light (around 700-800 nm):** While not as efficiently absorbed by chlorophyll, far-red light interacts with other photoreceptors like phytochromes, influencing processes like germination, stem elongation, and flowering. The ratio of red to far-red light can be a signal of shading by other plants.
  - **Ultraviolet (UV) Light (below 400 nm):** High doses of UV radiation can be damaging to plant DNA and photosynthetic machinery.

However, some UV-B can also trigger protective mechanisms in plants and influence the production of certain compounds.

- **Photomorphogenesis: Light as a Developmental Signal**

Beyond its role in energy provision, light acts as a critical environmental signal that influences plant growth and development – a process called **photomorphogenesis**. Plants have specialized photoreceptor proteins that detect different wavelengths of light and trigger various developmental responses.

- **Phytochromes:** These photoreceptors are sensitive to red and far-red light. They play a crucial role in:
    - **Germination:** In some seeds, red light promotes germination, while far-red light inhibits it. This allows seeds to sense if they are under a canopy (high far-red) or in open light (high red).
    - **Shade Avoidance:** A low red to far-red ratio (indicating shade) triggers stem elongation, reduced branching, and upward leaf movement to try and reach more light.
    - **Flowering (Photoperiodism):** Phytochromes are central to the plant's ability to measure day length and initiate flowering in response to seasonal changes.
- **Cryptochromes and Phototropins:** These photoreceptors are sensitive to blue and UV-A light. They are involved in:
    - **Phototropism:** The growth of a plant towards a light source.
    - **Stomatal Opening:** Blue light stimulates the opening of stomata, facilitating gas exchange for photosynthesis.
    - **Circadian Rhythms:** Regulating the plant's internal daily clock.
- **Photoperiodism (Detailed):** As mentioned earlier, the duration of the light period (day length) is a critical cue for flowering in many plants.
    - **Short-day plants (SDPs):** Flower when the period of darkness exceeds a critical length. Examples include chrysanthemums, poinsettias, and soybeans. It's actually the length of the uninterrupted dark period that is crucial.
    - **Long-day plants (LDPs):** Flower when the period of light exceeds a critical length (or the dark period is shorter than a critical length). Examples include spinach, lettuce, and wheat.
    - **Day-neutral plants (DNPs):** Flowering is not significantly influenced by day length. Examples include tomatoes, cucumbers, and some varieties of peppers.

5.7.4. Impact on Crop Yield and Quality

Ultimately, the effect of light translates into the yield and quality of the harvested crop:

- **Biomass Accumulation:** Sufficient light for photosynthesis directly drives the production of plant biomass (leaves, stems, roots, fruits, grains). Higher light intensity (up to the saturation point) generally leads to greater biomass.
- **Reproductive Development:** The right light duration and quality are essential for timely and successful flowering, pollination, and fruit/seed development. Insufficient or inappropriate light can delay or prevent flowering, reduce fruit set, and affect seed fill.
- **Nutritional Content:** Light can influence the levels of certain nutrients, vitamins, and secondary metabolites in crops. For example, higher light intensity can sometimes increase the concentration of certain antioxidants.
- **Morphology and Architecture:** Optimal light conditions lead to well-structured plants that can efficiently capture light and support fruit/grain development. Poor light can result in leggy, weak plants with reduced yield.

5.7.5. In Practical Terms for Crop Production

- **Field Layout:** Farmers consider the orientation and spacing of rows to maximize light interception by the plant canopy.
- **Pruning and Training:** Techniques to improve light penetration within the canopy.
- **Supplemental Lighting:** In greenhouses, artificial lighting (often LEDs with specific light spectra) is used to extend day length or increase light intensity, especially during winter or in cloudy regions.
- **Shading:** In very hot and sunny climates, shading can be used to prevent photoinhibition and reduce water stress.

Understanding the intricate relationship between light and plant physiology is fundamental to optimizing crop production for both quantity and quality.

6

Atmospheric Temperature

Atmospheric temperature refers to the **measure of the warmth or coldness of the air at different levels within the Earth's atmosphere.** It's not a uniform value but varies significantly with altitude and location.

More detailed descrption of atmospheric temperature is as follows:

- **Different Layers, Different Temperatures:** The Earth's atmosphere is divided into several layers (troposphere, stratosphere, mesosphere, thermosphere, and exosphere), and each layer has a distinct temperature profile. For example, the troposphere (where we live and where weather occurs) generally gets colder with increasing altitude, while the stratosphere above it gets warmer due to the absorption of ultraviolet radiation by the ozone layer.
- **Surface Temperature:** When people talk about the "temperature" in daily life, they usually refer to the temperature of the air near the Earth's surface, which is measured at meteorological stations.
- **Global Average:** Climatologists also use the concept of a "global temperature," which is an average of temperature measurements taken at the surface, near-surface, or in the troposphere across the entire planet.

6.1. Factors Affecting Atmospheric Temperature

The temperature of the atmosphere at any given point is influenced by a multitude of factors, including:

- **Solar Radiation:** The primary source of heat for the Earth and its atmosphere. The amount of solar radiation received varies with latitude, season, and time of day.
- **Latitude:** Regions closer to the equator receive more direct sunlight and are generally warmer than those at higher latitudes.
- **Altitude:** In the troposphere, temperature generally decreases with increasing altitude.
- **Proximity to Large Bodies of Water:** Water has a moderating effect on temperature. Coastal areas tend to have smaller temperature ranges compared to inland areas.

- **Ocean Currents:** Warm and cold ocean currents can influence the temperature of the air above them and nearby land.
- **Air Masses:** The movement of large bodies of air with different temperature and humidity characteristics affects local temperatures.
- **Land Surface:** Different surfaces (like forests, deserts, ice) absorb and reflect solar radiation differently, influencing the temperature of the air above them.
- **Cloud Cover:** Clouds can reflect incoming solar radiation (cooling effect during the day) and trap outgoing infrared radiation (warming effect at night).
- **Humidity:** Higher humidity can make the air feel warmer and can influence how quickly the air temperature changes.

6.2. Diurnal Variation of Temperature

The **diurnal variation of temperature** refers to the pattern of temperature change that occurs over a 24-hour period (a day-night cycle). It describes how the temperature typically rises during the day, reaches a peak in the afternoon, and then falls during the night to a minimum before sunrise.

6.2.1. Key Aspects of Diurnal Temperature Variation

- **Daily Cycle:** It's a recurring pattern that happens every day due to the Earth's rotation and the resulting changes in solar radiation.
- **Maximum Temperature:** Usually occurs in the afternoon, not exactly at noon when the sun is highest, because the Earth's surface and the air above it continue to absorb heat for some time after.
- **Minimum Temperature:** Typically occurs around dawn, as the Earth continues to lose heat throughout the night.
- **Diurnal Temperature Range (DTR):** This is the difference between the maximum and minimum temperatures recorded during a 24-hour period. A large DTR indicates a significant swing in temperature between day and night, while a small DTR means the temperature remains relatively stable.

6.2.2. Factors influencing the diurnal variation of temperature

- **Solar Radiation:** The primary driver. More sunlight leads to higher temperatures.
- **Cloud Cover:** Clouds can reduce the amount of solar radiation reaching the surface during the day (leading to cooler temperatures) and trap outgoing heat at night (leading to warmer nighttime temperatures), thus reducing the DTR.

- **Humidity:** High humidity can moderate temperature changes because water vapor in the air can absorb and retain heat, leading to a smaller DTR. Dry air allows for more rapid heating during the day and cooling at night, resulting in a larger DTR.
- **Surface Type:** Different surfaces (e.g., water, land, vegetation) heat up and cool down at different rates. For example, land heats and cools faster than water.
- **Wind:** Wind can mix the air, reducing temperature extremes near the surface.
- **Latitude:** Locations closer to the equator tend to have less diurnal variation compared to higher latitudes.

  Altitude: Higher altitudes can experience greater temperature drops at night due to thinner air.
- **Proximity to Water Bodies:** Large bodies of water moderate temperature swings, leading to smaller diurnal variations in coastal areas.

6.3. Lapse Rate

The **lapse rate** is the rate at which an atmospheric variable, most commonly **temperature**, decreases with an increase in altitude (Fig. 6.1). It's usually expressed in degrees Celsius (or Fahrenheit) per kilometer (or per 1000 feet).

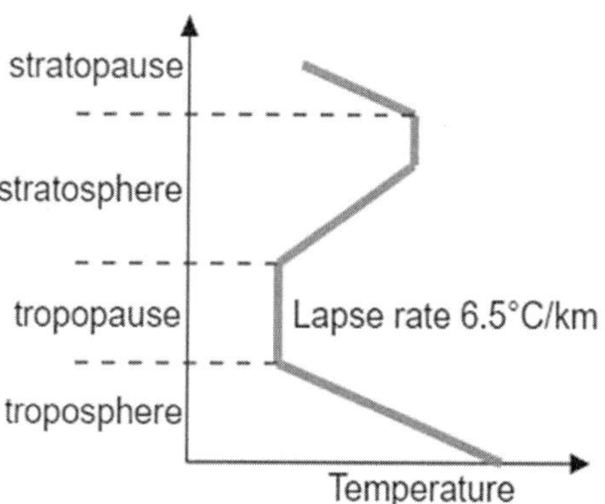

Fig. 6.1: Lapse rate

There are different types of lapse rates

- **Environmental Lapse Rate (ELR):** This is the actual rate of temperature decrease with altitude at a specific time and location. It's what you would measure with a weather balloon. The ELR is not constant and varies depending on weather conditions.

- **Dry Adiabatic Lapse Rate (DALR):** This is the rate at which a dry (unsaturated) parcel of air cools as it rises and expands adiabatically (without exchanging heat with its surroundings). The DALR is approximately 9.8°C per kilometer.
- **Moist Adiabatic Lapse Rate (MALR) or Saturated Adiabatic Lapse Rate (SALR):** This is the rate at which a saturated parcel of air cools as it rises. Because condensation releases latent heat, the MALR is less than the DALR and varies with temperature and pressure (typically between 4°C and 9°C per kilometer in the lower atmosphere).

6.3.1. Why Temperature Decreases with an Increase in Altitude (in the Troposphere)

The primary reason why temperature generally decreases with increasing altitude in the **troposphere** (the lowest layer of the atmosphere, where we live and where most weather occurs) is because the troposphere is primarily heated from the **ground up**, rather than directly by solar radiation.Here's a more detailed explanation:

- **Earth's Surface Absorption:** The Earth's surface absorbs solar radiation and warms up.
- **Heat Transfer to the Air:** This warmed surface then heats the air directly above it through:
    - **Conduction:** Direct transfer of heat through contact.
    - **Convection:** Warmer, less dense air rises, carrying heat upwards.
- **Expansion and Cooling of Rising Air:** As warm air rises, it moves into areas of lower atmospheric pressure. This lower pressure allows the air to expand. When a gas expands without an external heat source, it cools down (an adiabatic process). This is a significant reason for the temperature decrease with altitude.
- **Decreasing Air Density:** Higher altitudes have lower air density. Fewer air molecules mean less capacity to retain heat. While this is a contributing factor, the adiabatic cooling of rising air is the dominant reason for the lapse rate in the troposphere.

It may be summarized that the air closest to the Earth's surface is heated most effectively. As this warm air rises, it expands and cools, leading to a decrease in temperature with increasing altitude in the troposphere. The lapse rate quantifies this decrease.

Keep in mind that in other layers of the atmosphere (like the stratosphere and thermosphere), the temperature profile is different due to different mechanisms of energy absorption (e.g., ozone absorbing UV radiation in the stratosphere). However, the question specifically asks about the general decrease with altitude, which applies primarily to the troposphere.

6.4. Temperature Inversion

- Temperature inversion occurs when warm air forms over cool air, creating a "cap" that prevents pollutants and moisture from rising (Fig. 6.2). This leads to poor air quality, as pollutants accumulate and cause respiratory issues and other health problems. It also results in fog and haze, reducing visibility and affecting transportation. Additionally, temperature inversions can lead to unusual weather patterns, such as frosts, freezes, or altered precipitation patterns.

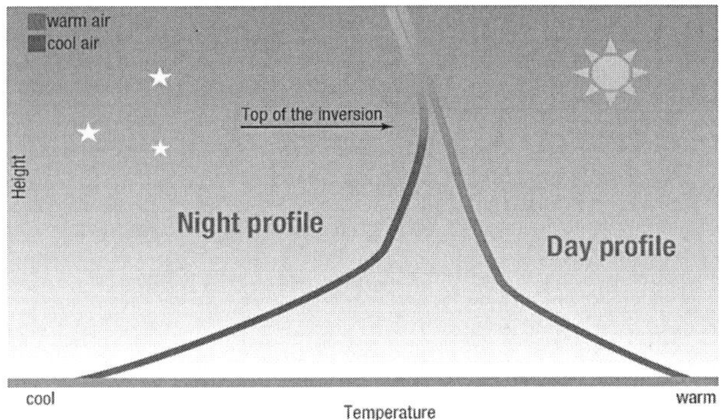

Fig. 6.2: Temperature inversion

- Understanding temperature inversions is crucial for environmental monitoring, allowing for tracking of air quality and pollution levels. It's also essential for weather forecasting, enabling predictions of weather patterns and warning systems. Furthermore, recognizing temperature inversions helps mitigate health issues related to poor air quality, making it a vital aspect of public health.
- Temperature inversions can impact various aspects of life, including aviation, transportation, and daily activities. In areas prone to temperature inversions, such as valleys or urban areas, monitoring and understanding this phenomenon is essential for managing air quality, predicting weather, and protecting public health.

6.5. Vertical Profile of Air Temperature in the Atmosphere

The vertical profile of air temperature in the atmosphere is complex and varies significantly with altitude. The atmosphere is divided into several layers based on these temperature variations (Fig. 6.2).

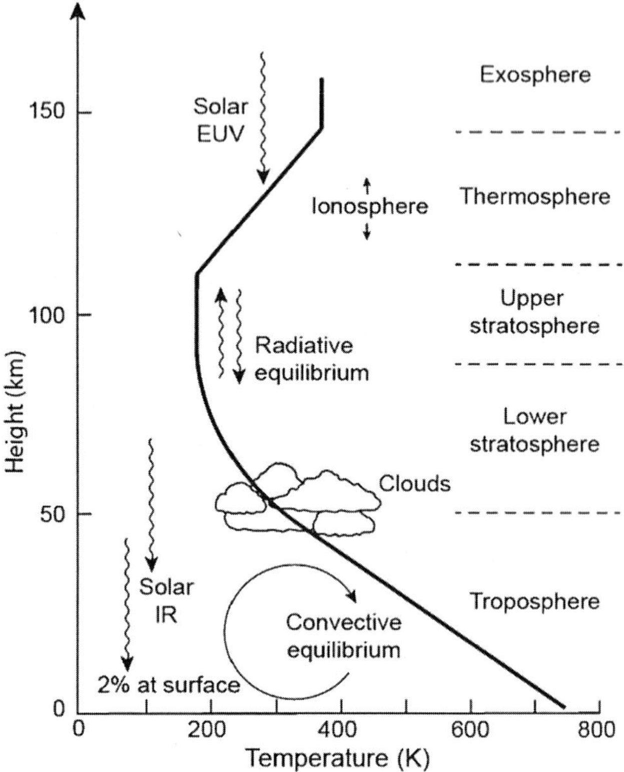

Fig. 6.3: Vertical profile of air temperature in the atmosphere

The detailed description is as under:

6.5.1. Troposphere

- **Altitude:** Extends from the Earth's surface up to an average of 10-15 km (it's thicker at the equator and thinner at the poles).
- **Temperature Profile:** Temperature generally **decreases with altitude**. This is because the Earth's surface absorbs most of the solar radiation and heats the air from below. As you move higher, you get further away from this heat source.
- **Lapse Rate:** The average rate of temperature decrease is about 6.5°C per kilometer, known as the environmental lapse rate.
- **Key Features:** Most weather phenomena (clouds, rain, snow) occur in this layer. It contains about 75-80% of the atmosphere's mass and almost all of its water vapor. The top of the troposphere is called the **tropopause**.

6.5.2. Stratosphere

- **Altitude:** Extends from the tropopause to about 50 km.
- **Temperature Profile:** Temperature **increases with altitude** in this layer
- **Reason for Increase:** This warming is due to the absorption of ultraviolet (UV) radiation from the Sun by the **ozone layer**, which is located within the stratosphere. This absorption converts UV energy into heat.
- **Key Features:** The stratosphere is more stable than the troposphere because the warmer air is above the cooler air, inhibiting vertical mixing. The top of the stratosphere is called the **stratopause**. Temperatures range from around −60°C at the tropopause to about −15°C at the stratopause.

6.5.3. Mesosphere

- **Altitude:** Extends from the stratopause to about 85 km.
- **Temperature Profile:** Temperature **decreases with altitude** again in the mesosphere.
- **Reason for Decrease:** There is no significant absorption of solar radiation in this layer to cause warming. Radiative cooling dominates.
- **Key Features:** The mesosphere is the coldest part of the Earth's atmosphere, with temperatures reaching as low as −90°C near the top (the **mesopause**). Meteors burn up in this layer.

6.5.4. Thermosphere

- **Altitude:** Extends from the mesopause outwards, with no well-defined upper limit (some sources say it starts around 80-90 km and goes up to 500-1000 km).
- **Temperature Profile:** Temperature **increases dramatically with altitude**.
- **Reason for Increase:** Absorption of high-energy solar radiation (like X-rays and UV) by the sparse gases in this layer causes this temperature increase.
- **Key Features:** Despite the high temperatures (which can reach hundreds or even thousands of degrees Celsius), it would feel very cold to us because the air density is extremely low, meaning there are very few molecules to transfer heat. The **ionosphere**, a region where gases are ionized by solar radiation, overlaps with the thermosphere. Auroras (Northern and Southern Lights) occur in this layer.

6.5.5. Exosphere

- **Altitude:** The outermost layer of the atmosphere, starting from the thermopause and gradually fading into space.
- **Temperature Profile:** Temperatures are very high here, but the concept of temperature is less meaningful due to the extremely low density.
- **Key Features:** Gases are very sparse, and atoms and molecules can escape into space. Satellites also orbit in this region.

In summary, the vertical profile of air temperature in the Earth's atmosphere is characterized by alternating layers of decreasing and increasing temperatures with altitude, primarily due to the way solar radiation is absorbed at different levels and the changing density of the air.

6.6. Seasonal Variation in Temperature

Seasonal variation in temperature refers to the cyclical changes in air temperature that occur throughout the year. These variations are primarily driven by the Earth's revolution around the Sun and the tilt of the Earth's axis. The key points are:

- **Cause:** The Earth's axis is tilted at approximately 23.5 degrees relative to its orbital plane. As the Earth revolves around the Sun, this tilt causes different parts of the Earth to receive more direct sunlight at different times of the year.
- Direct vs. Oblique Sunlight
    - When a hemisphere is tilted towards the Sun, it receives more direct sunlight. Direct rays are more concentrated, leading to higher temperatures (summer).
    - When a hemisphere is tilted away from the Sun, it receives more oblique (angled) sunlight. Oblique rays spread the solar energy over a larger area and pass through more of the atmosphere, resulting in lower temperatures (winter).
- **Northern and Southern Hemispheres:** The seasons in the Northern and Southern Hemispheres are opposite. When it is summer in the Northern Hemisphere (around June solstice), it is winter in the Southern Hemisphere, and vice versa (around December solstice).
- **Equinoxes:** During the spring (vernal) and autumn (autumnal) equinoxes (around March 20th and September 22nd/23rd), the Earth's tilt is neither towards nor away from the Sun. At these times, most places on Earth experience roughly 12 hours of daylight and 12 hours of night.

- Solstices
  - **Summer Solstice (around June 21st in the Northern Hemisphere):** The Northern Hemisphere is tilted most directly towards the Sun, resulting in the longest day of the year in the Northern Hemisphere and the shortest in the Southern Hemisphere.
  - **Winter Solstice (around December 21st in the Northern Hemisphere):** The Northern Hemisphere is tilted farthest away from the Sun, resulting in the shortest day of the year in the Northern Hemisphere and the longest in the Southern Hemisphere.
- **Latitude Dependence:** The magnitude of seasonal temperature variation increases as you move away from the equator towards the poles. The tropics experience relatively little seasonal temperature change, while the polar regions have the most extreme variations.
- **Other Influencing Factors:** While the Earth's tilt and orbit are the primary causes, other factors can modify seasonal temperature variations, including:
  - **Proximity to large bodies of water:** Water moderates temperature, leading to smaller seasonal swings in coastal areas compared to inland areas.
  - **Ocean currents:** Warm and cold ocean currents can influence the temperature of adjacent landmasses.
  - **Altitude:** Higher altitudes generally experience cooler temperatures throughout the year, and this can affect the seasonal range.
  - **Prevailing winds:** Winds can transport air masses with different temperatures, influencing local seasonal patterns.

6.7. Annual Variation in Temperature

Annual variation in temperature refers to the overall pattern of temperature changes that occur throughout a complete year at a particular location. It encompasses the progression from the lowest temperatures of winter through the warmth of summer and back again. This variation is best understood by looking at the average temperatures for each month of the year. The key aspects are:

- **Overall Yearly Trend:** It describes the general swing in temperature over twelve months, highlighting the warmest and coldest periods.
- **Monthly Averages:** Annual variation is often represented by the average temperature of each month, showing how the typical temperature changes as the year progresses.

- **Annual Temperature Range:** A key metric derived from the annual variation is the **annual temperature range**. This is the difference between the average temperature of the warmest month and the average temperature of the coldest month. A large annual range indicates significant differences between summer and winter temperatures, while a small range suggests a more consistent temperature throughout the year.

6.8. Parameters Used for Description of Temperature

The meanings of these temperature terms are as follows:

- **Daily Maximum Temperature:** This is the **highest temperature recorded** during a 24-hour period (typically from midnight to midnight, or based on the observation schedule of a weather station). For example, the daily maximum temperature in Delhi today (May 25, 2025) is forecast to be 38°C.
- **Daily Minimum Temperature:** This is the **lowest temperature recorded** during a 24-hour period (corresponding to the same period as the daily maximum). For Delhi today, the forecast low is 28°C.
- **Diurnal Range of Temperature:** This is the **difference between the daily maximum temperature and the daily minimum temperature**. It indicates the temperature swing that occurs over a single day. For example, if the daily maximum is 38°C and the daily minimum is 28°C, the diurnal range is 38−28=10°C.
- **Mean Monthly Maximum and Minimum Temperatures:** To get the **mean monthly maximum temperature**, you would take the daily maximum temperatures for every day of a particular month, sum them up, and then divide by the number of days in that month. This gives you the average of the highest temperatures experienced each day within that month. Similarly, the **mean monthly minimum temperature** is calculated by taking the daily minimum temperatures for every day of the month, summing them, and dividing by the number of days. This gives the average of the lowest temperatures experienced each day within that month.
- **Mean Seasonal Maximum and Minimum Temperatures:** Seasons are typically defined as groups of months (e.g., Summer: June, July, August). The **mean seasonal maximum temperature** is found by averaging all the daily maximum temperatures within that entire season and then dividing by the total number of days in the season. The **mean seasonal minimum temperature** is calculated similarly, but using all the daily minimum temperatures within the season.

- **Mean Annual Maximum and Minimum Temperature:** The **mean annual maximum temperature** is calculated by averaging the daily maximum temperatures for all 365 (or 366 in a leap year) days of the year. The **mean annual minimum temperature** is the average of all the daily minimum temperatures throughout the entire year.
- **Annual Range in Temperature:** This is the **difference between the mean temperature of the warmest month and the mean temperature of the coldest month** of the year. It indicates the overall temperature variation throughout the year at a location. Note that this is different from simply the difference between the absolute highest and lowest temperatures recorded in a year.

In summary

- **Daily** terms refer to a 24-hour period.
- **Monthly** terms are averages over a calendar month.
- **Seasonal** terms are averages over a defined season (several months).
- **Annual** terms are averages over a whole year.
- The **range** is the difference between the maximum and minimum values of a set.

6.9. Isotherms

An **isotherm** (from the Greek "iso" meaning equal and "therm" meaning heat) is a line drawn on a map or chart that connects points having the **same temperature** at a given time or the same mean temperature for a given period (Fig. 6.3). Think of it like contour lines on a topographic map, but instead of showing points of equal elevation, isotherms show points of equal temperature.

Key characteristics of isotherms

- They represent a specific temperature value.
- Points along a single isotherm have the same temperature.
- The spacing between isotherms indicates the temperature gradient:
  - Isotherms that are close together indicate a rapid change in temperature over a short distance.
  - Isotherms that are far apart indicate a gradual change in temperature.
- Isotherms generally run east to west because temperature is largely controlled by latitude (the amount of solar radiation received). However, they can be modified by other factors like altitude, land and water distribution, and ocean currents.

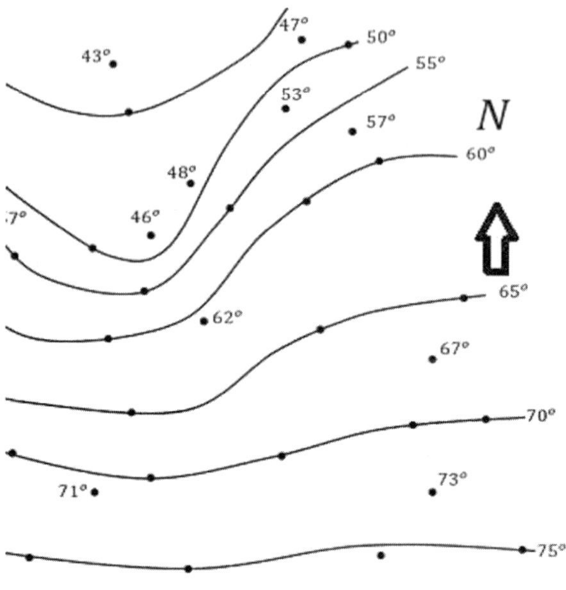

Fig. 6.4: Isotherms

6.9.1. Isothermal Chart

An isothermal chart is a map that displays isotherms, allowing for the visualization of temperature distribution over a geographical area. These charts are used to:

- Show the spatial patterns of temperature.
- Identify areas with similar temperatures.
- Analyze temperature gradients.
- Compare temperatures across different regions.
- Track changes in temperature over time (if a series of charts are used).

Isothermal charts can depict various temperature data, such as

- **Surface air temperature:** Showing the temperature near the ground at a specific time.
- **Mean monthly temperature:** Illustrating the average temperature for a particular month across a region.
- **Mean annual temperature:** Showing the average temperature over a whole year.

Therefore, an isothermal chart uses isotherms to provide a visual representation of how temperature varies geographically.

6.10. Temperature Conversion

It is a well-known fact in temperature conversions that **– 40°F = – 40°C**

By applying following temperature conversion formulae,

Fahrenheit to Celsius: $C = (F-32) \times 5/9$

Celsius to Fahrenheit: $F = C \times 9/5 + 32$,

it can be proved that **– 40°F = – 40°C**

7

Thermal Regime: Crop Growth and Development

Temperature stands as a fundamental environmental factor profoundly influencing the life cycle of crops, from the initial spark of germination to the final stages of maturation and yield formation. The **thermal regime**, encompassing the daily and seasonal patterns of temperature, dictates the rate of biochemical reactions within the plant, thereby governing the pace and efficiency of physiological processes crucial for growth and development.

This chapter, „Thermal Regime: Crop Growth and Development," will delve into the intricate relationship between temperature and the various phenological stages of crop plants. It will be explored how different cardinal temperatures – the minimum, optimum, and maximum thresholds – dictate the boundaries within which growth can occur and the rates at which developmental milestones are achieved. Understanding the impact of temperature on processes like photosynthesis, respiration, nutrient uptake, and ultimately, yield, is paramount for optimizing crop production in diverse environments.

Furthermore, we will examine the concepts of heat units or growing degree days, which provide a quantitative framework for predicting crop development based on temperature accumulation. By appreciating how the thermal environment interacts with genetic potential and other environmental factors, we can gain valuable insights into crop suitability for specific regions, the timing of critical growth stages, and strategies for mitigating the adverse effects of temperature extremes.

This chapter aims to provide a comprehensive understanding of the thermal regime and its critical role in shaping the growth and development of our cultivated plants, laying the foundation for informed agricultural practices and improved crop management.

7.1. Crop Growth

Crop growth refers to the **irreversible increase in the size or mass of a plant or its parts over time.** This increase can be measured in various ways, such as:

- **Height:** Increase in the vertical length of the stem.
- **Leaf area:** Expansion of the surface area of the leaves.
- **Biomass (dry weight):** The total dry matter accumulated by the plant.
- **Volume:** Increase in the overall bulk of the plant or its organs.

7.1.1. Explanation

Crop growth is a fundamental process in agriculture, representing the accumulation of biological material through various physiological activities. It's not just about getting taller; it encompasses the development of all plant organs – roots, stems, leaves, and eventually reproductive structures.

7.1.2. Key Aspects of Crop Growth

- **Irreversible:** Unlike changes due to water uptake (which can be reversed), true growth involves the creation of new cells and the expansion of existing ones, leading to a permanent increase in size or mass.
- **Metabolic process:** Growth is driven by metabolic activities within the plant, primarily photosynthesis, where light energy is converted into chemical energy, fueling the production of new organic matter. Respiration provides the energy for these processes.
- **Influenced by many factors:** Crop growth is not a standalone process. It's heavily influenced by both **internal (genetic)** factors and **external (environmental)** factors.
  - **Internal factors:** The genetic makeup of the crop variety dictates its inherent growth potential, rate of development, and final size.
  - **External factors** include:
  - **Light:** Essential for photosynthesis. Quantity, quality, and duration all play a role.
  - **Temperature:** Affects the rate of metabolic reactions. Each crop has optimal, minimum, and maximum temperatures for growth.
  - **Water:** Crucial for cell turgidity, nutrient transport, and photosynthesis.
  - **Nutrients:** Macronutrients (N, P, K, Ca, Mg, S) and micronutrients (e.g., Fe, Mn, Zn, Cu, B, Mo) are essential building blocks and cofactors for plant processes.
  - **Air:** Carbon dioxide for photosynthesis and oxygen for respiration.
  - **Humidity:** Affects transpiration rates.
  - **Soil:** Provides anchorage, water, and nutrients.
  - **Biotic factors:** Pests, diseases, and beneficial organisms.

In essence, crop growth is the quantitative increase in the plant's physical dimensions and biological mass over time, resulting from the complex interplay of its genetic potential and the surrounding environment. Understanding crop growth is vital for optimizing agricultural practices to achieve high yields.

7.2. Crop Growth Rate

Crop growth rate (CGR) is a term used in agriculture and botany to quantify how quickly a crop is growing. It measures the increase in plant biomass (dry matter) per unit of land area over a specific period.

7.2.1. What it Measures?

- CGR specifically focuses on the accumulation of dry matter, which represents the organic material produced by the plant through photosynthesis (Fig. 7.1). This is a more accurate measure of growth than simply measuring height, as plants can increase in height without necessarily increasing in biomass.

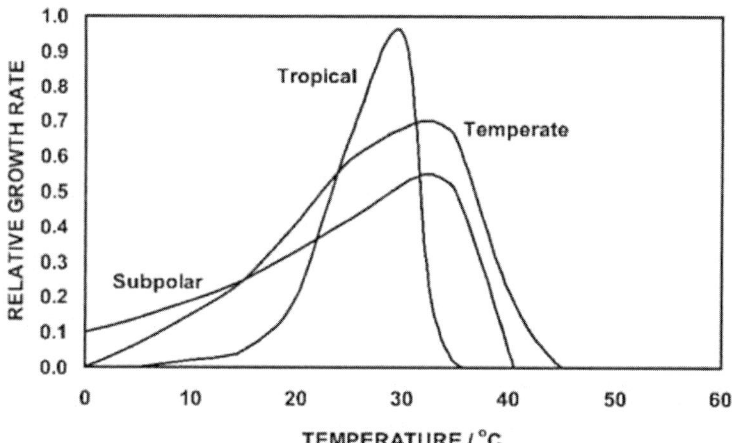

Fig. 7.1: Crop growth rate vs. temperature

The units for CGR are typically expressed as grams per square meter per day (g m^{-2} day^{-1}). This indicates how many grams of new plant material are being produced on each square meter of land each day.

7.2.2. Why it's Important?

- CGR is a key indicator of crop productivity and overall plant health.
- It helps in assessing the effectiveness of agricultural practices, such as fertilization, irrigation, and crop management.
- Farmers and researchers use CGR to:

- Compare the growth rates of different crop varieties.
- Evaluate the impact of environmental factors (e.g., temperature, light, water) on crop growth.
- Optimize crop management practices to maximize yield.
- Predict crop yields.

7.2.3. Factors Affecting CGR

Several factors influence how quickly a crop grows:

- **Photosynthesis:** The rate at which the plant converts sunlight, carbon dioxide, and water into organic matter.
- **Leaf area index (LAI):** The amount of leaf area per unit of land area. A higher LAI generally means more photosynthesis and a higher CGR, up to a certain point.
- **Temperature:** Plants have an optimal temperature range for growth.
- **Water availability:** Adequate water is essential for photosynthesis and other physiological processes.
- **Nutrient availability:** Plants need sufficient nutrients (nitrogen, phosphorus, potassium, etc.) for healthy growth.
- **Crop variety:** Different crop species and varieties have different inherent growth rates.
- **Plant density:** The number of plants per unit area can affect CGR.
- **Growth stage:** CGR varies throughout the plant‘s life cycle, typically peaking during the vegetative stage and declining as the plant reaches maturity.

In essence, crop growth rate provides a quantitative measure of how efficiently a crop is converting resources into biomass, making it a valuable tool for understanding and managing crop production.

7.3. Crop Development

Crop development refers to the changes in the plant‘s form and function as it progresses through its life cycle (qualitative change). Crop development involves the progression of a plant through various **phenological stages**, from germination to maturity. These stages include:

- **Germination:** The sprouting of a seed.
- **Vegetative stage:** The growth of stems, leaves, and roots.
- **Reproductive stage:** The development of flowers, fruits, and seeds.
- **Maturation:** The final stage where the plant reaches full development.
- **Key points about crop development**

- Development is a continuous process, with gradual transitions between stages.
- It is influenced by both genetic factors and environmental cues.
- Environmental factors like temperature and photoperiod (day length) play a significant role in regulating the timing of developmental stages. For instance, some plants require specific day lengths to initiate flowering.
- Development is characterized by changes in gene expression, hormone production, and physiological processes that lead to the formation of different plant structures and functions.

Understanding crop development is crucial in agriculture because it helps determine:

- The timing of critical growth stages.
- The optimal time for applying inputs like fertilizers and pesticides.
- The suitability of a crop to a particular environment.
- Strategies to improve crop yield and quality.

7.4. Heat Units

Heat units, also known as growing degree days (GDD), are a measure of heat accumulation over time. They are used in agriculture and biology to estimate the growth and development of plants, insects, and other organisms (Fig. 7.2).

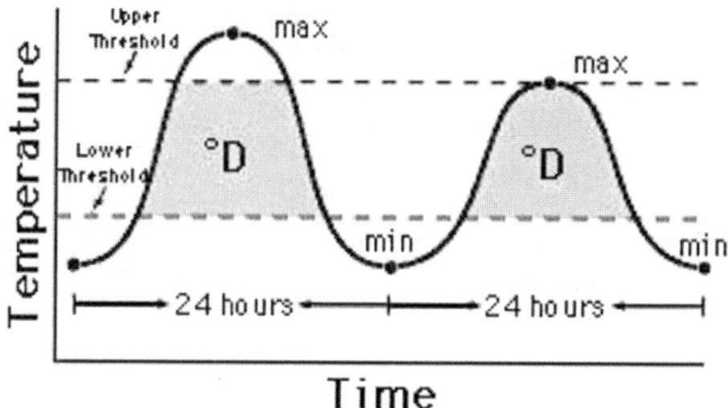

Fig. 7.2: Heat units or growing degree days

7.4.1. Definition

In simpler terms, heat units quantify the amount of warmth that accumulates, influencing the rate at which biological processes occur. Many biological organisms, like plants, develop within a specific temperature range. Heat units help us determine how quickly an organism progresses through its lifecycle, based on the temperature.

7.4.2. Method of Calculation

The most common method to calculate heat units involves these steps:

- **Determine the base temperature:** This is the minimum temperature at which the organism (e.g., a specific crop) begins to grow or develop. Temperatures below this base temperature are considered ineffective for development.
- **Obtain daily temperatures:** Record the daily maximum and minimum temperatures.
- **Calculate daily heat units**
  - For each day, the calculation is:
  - Daily Heat Units = [(Daily MaxTemp + Daily Min Temp) / 2] - Base Temperature
  - If the result is negative, the daily heat units are considered to be zero, as no development occurs below the base temperature.
  - If the maximum temperature exceeds a crop-specific upper threshold, it is often capped at that threshold for the calculation.
- **Accumulate heat units:** Sum the daily heat units from the start of the period of interest (e.g., planting date) to get the total heat units.

7.4.3. Formula

Daily Heat Units= [(Tmax + Tmin)/2] - Tbase

Where:

- Tmax = Daily maximum temperature
- Tmin = Daily minimum temperature
- Tbase = Base temperature
  - **Example:** Let's calculate the heat units for a day for a crop with a base temperature of 10°C (summer season & 5°C for winter season crop), with a maximum temperature of 25°C and a minimum temperature of 15°C.

Daily Heat Units = [(25 + 15)/2] - 10

= [40/2] - 10

= 20 - 10

= 10

Therefore, 10 heat units (day°C)would be accumulated for that day.

7.4.4. Importance of Heat Units

The concept of heat units, or growing degree days (GDD), is of great importance in agriculture and has several practical applications:

- **Predicting Plant Development**
    - Different plants require a specific amount of heat units to reach various growth stages, such as germination, flowering, and maturity.
    - Farmers can use this information to predict when crops will be ready for harvest, allowing for better planning of harvesting activities.
- **Crop Selection and Zoning**
    - Heat unit requirements vary among different crop varieties.
    - By understanding the heat units available in a specific region, farmers can choose the most suitable varieties that will thrive in that climate.
    - This helps in optimizing crop production and minimizing the risk of crop failure.
- **Optimizing Planting Schedules**
    - Heat units help determine the optimal planting dates for different crops.
    - Planting at the right time ensures that crops receive the required heat units during their critical growth stages, leading to higher yields.
- **Pest Management**
    - The development of insects and other pests is also influenced by temperature.
    - Heat units can be used to predict pest life cycle stages, allowing farmers to time their pest control measures effectively.
- **Irrigation Scheduling**
    - By tracking heat units, farmers can estimate the water requirements of crops at different growth stages.
    - This information helps in scheduling irrigation, preventing water stress, and optimizing water use efficiency.
- **Climate Change Adaptation**
    - Heat units can be used to monitor the effects of climate change on crop development.
    - Changes in heat unit accumulation patterns can help farmers adapt their practices, such as shifting planting dates or selecting more heat-tolerant varieties.

In summary, heat units are a valuable tool for agricultural planning and decision-making. They help farmers optimize crop production, manage resources efficiently, and adapt to changing environmental conditions.

7.5. Heat and Cold Waves

The heat waves and cold waves indicate extremes of temperatures prevailing for extended periods and are injurious to crop growth.

- **Heat Wave:** A period of unusually hot and dry weather that lasts for several days.
- **Cold Wave:** A period of abnormally low temperatures that lasts for several days.
  - **Definitions in India:** The India Meteorological Department (IMD) defines heat wave and cold wave conditions based on temperature thresholds and deviations from normal temperatures.

7.5.1. Heat Wave

- A heat wave is declared when the maximum temperature reaches at least 40°C in the plains and 30°C in hilly regions (Fig. 7.3).

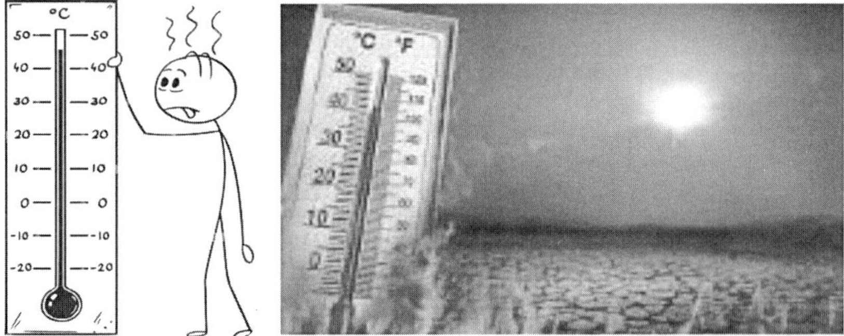

Fig. 7.3 Heat wave

- If the normal maximum temperature is less than or equal to 40°C, a heat wave is declared when the actual maximum temperature is 5°C to 6°C higher than the normal. A severe heat wave is when it‘s more than 7°C higher.
- If the normal maximum temperature is more than 40°C, a heat wave is declared when the actual maximum temperature is 4°C to 5°C higher than the normal. A severe heat wave is when it‘s more than 6°C higher.
- Additionally, if the actual maximum temperature remains 45°C or more, a heat wave is declared, irrespective of the normal maximum temperature.

7.5.2. Cold Wave

- A cold wave is declared when the minimum temperature is 10°C or below in the plains and 0°C or below in hilly regions (Fig. 7.4).

Fig. 7.4: Cold wave

- It is also defined based on the departure of the minimum temperature from the normal.
- **Effects on Crops:** Both heat waves and cold waves can have significant adverse effects on crop production:

7.5.3. Heat Waves

- **Reduced Yield:** High temperatures can cause crops to wilt, dry out, and experience reduced photosynthesis, leading to lower yields.
- **Crop Damage:** Extreme heat can damage plant tissues, affecting their growth and development.
- **Water Stress:** Heat waves increase evaporation, leading to water stress and drought-like conditions that harm crop growth.
- **Increased Pest and Disease:** Some pests and diseases thrive in hot conditions, potentially increasing their infestation and impact on crops.
- **Premature Ripening:** High temperatures can cause premature ripening, affecting the quality and quantity of the yield.
- **Impact on pollination:** Extreme heat can disrupt the pollination process in many crops, leading to lower fruit and seed production.

7.5.4. Cold Waves

- **Frost Damage:** Freezing temperatures can cause frost damage to plant cells, leading to wilting, tissue damage, and even plant death.
- **Delayed Development:** Cold waves can slow down or halt plant growth and development, delaying maturity and reducing yields.
- **Reduced Yield:** The damage caused by cold waves can result in significant yield losses for various crops.

- **Increased Vulnerability to Disease:** Cold stress can weaken plants, making them more susceptible to diseases.
- **Impact on flowering:** Low temperatures can affect the flowering of crops, which in turn affects the fertilization and yield.

7.6. Earths Surface Energy Budget

The Earth's surface receives energy primarily from solar radiation (Fig. 7.5). This energy is then redistributed through various processes, including:

- **Absorption:** The surface absorbs incoming solar radiation.
- **Reflection:** Some solar radiation is reflected back into the atmosphere.
- **Emission:** The surface emits thermal radiation (infrared).
- **Conduction:** Heat transfer through direct contact.
- **Convection:** Heat transfer through the movement of fluids (air and water).
- **Evaporation:** The process by which liquid water changes into vapor
- **Latent heat flux:** The energy absorbed or released during a phase transition (evaporation, condensation, etc.).
- **Sensible heat flux:** The transfer of energy that changes the temperature of the medium.
- The energy budget describes how these processes balance each other out.

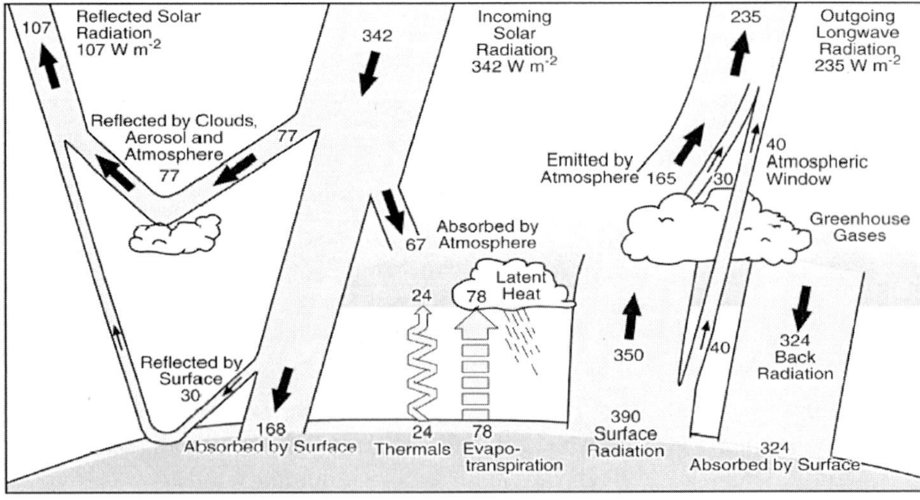

Fig. 7.5: Earth's energy balance

7.6.1. Equation

The surface energy budget can be represented by the following simplified equation:

R_net = H + LE + G + ΔS

Where:

The energy budget (Fig. 7.6) describes how these processes balance each other.

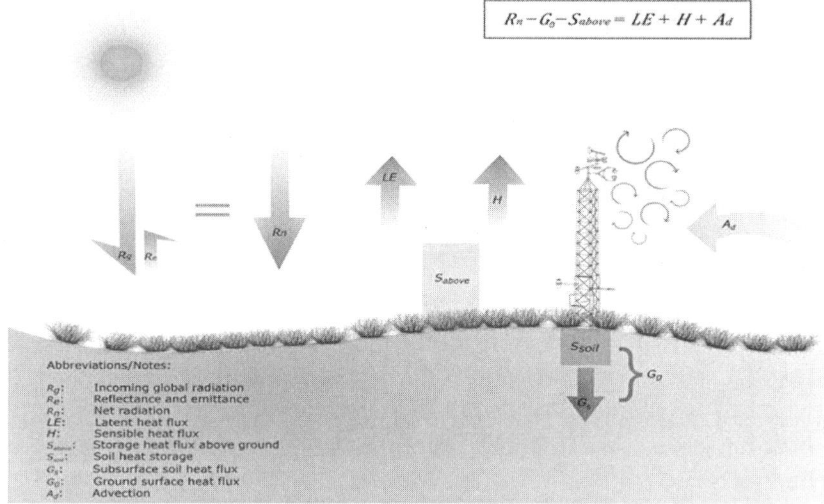

- **Equation :** The surface energy budget can be represented by the following simplified equation:
- **R_net** = (incoming solar radiation minus outgoing reflected and thermal radiation)
  - **H** is the sensible heat flux
  - **LE** is the latent heat flux
  - **G** is the ground heat flux
  - **ΔS** is the change in energy storage

A more detailed equation would be:

R_in - R_out = H + LE + G

Where:

- **R_in** is incoming radiation
- **R_out** is outgoing radiation (reflected and emitted)

7.6.2. Importance in Crop Production

The Earth's surface energy budget plays a crucial role in crop production in several ways:

- **Temperature Regulation:** The energy budget determines the temperature of the soil and air, which directly affects plant growth, development, and yield.
- **Photosynthesis:** Solar radiation, a key component of the energy budget, is the driving force behind photosynthesis, the process by which plants convert light energy into chemical energy for growth.
- **Evapotranspiration:** The balance between energy input and output influences evapotranspiration (the combined processes of evaporation and plant transpiration), which affects water availability for crops.
- **Microclimate:** The energy budget influences the microclimate around plants, including temperature, humidity, and airflow, which can affect crop health and productivity.
- **Crop Growth Stages:** Different stages of crop development have varying energy requirements. The energy budget helps determine whether these requirements are met. For example, during flowering and fruiting, crops need optimal temperatures, determined by the energy budget.
- **Yield and Quality:** Ultimately, the surface energy budget affects crop yield and quality. Optimal energy balance leads to healthy plant growth, efficient photosynthesis, and adequate water availability, resulting in higher and better-quality yields.
- **Agricultural Practices:** An understanding of the energy budget can inform agricultural practices, such as:
    - **Irrigation:** To manage water availability and evapotranspiration.
    - **Mulching:** To alter soil temperature and moisture.
    - **Crop selection:** Choosing crops that are suited to the local energy budget.
    - **Timing of planting and harvesting:** To coincide with optimal energy conditions.
    - **Design of agricultural systems:** To optimize light capture and heat dissipation.

8

Water Vapor in the Atmosphere

The presence of water in its various phases (gas, liquid, and solid) within the atmosphere plays a pivotal role in shaping our weather and climate.This chapter, „Water Vapor in the Atmosphere," will delve into the crucial concept of atmospheric humidity – the amount of water vapor in the air. Water, unique in its ability to exist in all three states at Earth's temperatures, undergoes phase changes that are fundamental to the formation of clouds, precipitation, and atmospheric stability. Understanding the sources of atmospheric moisture, the ways it is measured and expressed, and its influence on atmospheric processes is essential for comprehending a wide range of meteorological phenomena. We will explore concepts such as evaporation, condensation, humidity indices, and the role of moisture in energy transfer within the Earth-atmosphere system. Join us as we unravel the dynamics of this vital component of our atmospheric environment.

8.1. Expressions Used for Water Vapor in the Atmosphere

The concepts of vapor pressure, saturated vapor pressure, vapor pressure deficit, and dew point has been discussed with detailed explanations:

8.1.1. Vapor Pressure (e)

Vapor pressure is the partial pressure exerted by water vapor molecules in the air. It represents the contribution of water vapor to the total atmospheric pressure. Imagine air as a mixture of gases; the vapor pressure is the „pressure" exerted solely by the water vapor component of that mixture.

- **Detailed Explanation**
  - Water molecules in liquid or solid form (like the surface of a lake or an ice crystal) are constantly in motion.1 Some of these molecules gain enough kinetic energy to escape into the gaseous phase (evaporation or sublimation).
  - These water vapor molecules then exert their own pressure as they move and collide with surfaces around them. This pressure is the vapor pressure.

- The amount of water vapor in the air, and consequently the vapor pressure, depends on factors like the availability of water for evaporation/sublimation and the ongoing processes of evaporation/ sublimation versus condensation/deposition.
- Vapor pressure is usually measured in units of pressure like Pascals (Pa), kilopascals (kPa), or hectopascals (hPa).

8.1.2. Saturated Vapor Pressure (es)

Saturated vapor pressure is the maximum possible partial pressure of water vapor that can exist in the air at a given temperature. It represents the point at which the air is holding the maximum amount of water vapor it can before condensation (or deposition) begins.

- **Detailed Explanation**
  - At any given temperature, there's a limit to how much water vapor the air can „hold." This limit is determined by the energy of the air molecules. Warmer air molecules have more energy and can prevent water vapor molecules from easily condensing back into liquid water, thus allowing more water vapor to exist in the gaseous phase.
  - When the rate of evaporation (or sublimation) equals the rate of condensation (or deposition), the air is said to be saturated, and the pressure exerted by the water vapor at this equilibrium is the saturated vapor pressure.
  - Saturated vapor pressure is solely a function of temperature. As temperature increases, the saturated vapor pressure also increases non-linearly. This means warmer air can hold significantly more water vapor at saturation than colder air.
  - You can think of saturated vapor pressure as the „capacity" of the air to hold water vapor at a specific temperature.
  - When the actual vapor pressure equals the saturated vapor pressure (e=es), the relative humidity is 100%, and the air is saturated.

8.1.3. Vapor Pressure Deficit (VPD)

Vapor Pressure Deficit (VPD) is the difference between the saturated vapor pressure (es) and the actual vapor pressure (e) at a given temperature.

VPD=es − e

- **Detailed Explanation**
  - VPD essentially indicates how „thirsty" the air is. A high VPD means the air is far from being saturated and has a high capacity to

hold more water vapor, leading to increased evaporation rates from surfaces (like plants, soil, and water bodies).

- A low VPD means the air is closer to saturation, and the driving force for evaporation is smaller. When VPD is zero (es =e), the air is saturated, and the net evaporation rate is zero (equilibrium).
- VPD is important in various fields, including agriculture (plant transpiration), ecology (evaporation rates), and even wildfire risk assessment (fuel dryness).
- It's often considered a more direct measure of the evaporative demand of the atmosphere than relative humidity because it directly relates to the pressure difference driving moisture movement.

8.1.4. Dew Point (Td)

The dew point is the temperature to which air must be cooled at a constant pressure and constant water vapor content for saturation to occur. At this temperature, the air can no longer hold all of its water vapor, and condensation begins to form dew, fog, or clouds.

- **Detailed Explanation**
  - Imagine a parcel of air with a certain amount of water vapor (and thus a specific vapor pressure, e). If you cool this air at a constant pressure without adding or removing any water vapor, the saturated vapor pressure (es) of the air decreases because es is temperature-dependent.
  - The dew point is the temperature at which this cooling causes the saturated vapor pressure (es) to become equal to the actual vapor pressure (e) of the air. At this point, the air becomes saturated (100 % relative humidity), and further cooling will lead to condensation.
  - A high dew point indicates a large amount of moisture in the air because the air needs to be cooled significantly less to reach saturation. A low dew point indicates drier air.
  - The difference between the air temperature and the dew point is a good indicator of how close the air is to saturation. A small difference means the relative humidity is high, and the air is close to saturation. A large difference means the relative humidity is low, and the air is dry.
- **In summary, the**
  - **Vapor Pressure** tells us how much water vapor is currently in the air.
  - **Saturated Vapor Pressure** tells us the maximum amount of water vapor the air *could* hold at the current temperature.

- **Vapor Pressure Deficit** tells us the „drying potential" of the air.
- **Dew Point** tells us the temperature at which condensation would begin if the air were cooled.

8.2. Concept of Saturation of Air

Here's what saturation indicates:

- **Maximum Water Vapor Capacity:** Saturated air has reached its limit for holding water vapor at its current temperature. Warmer air can hold more water vapor at saturation than colder air.
- **Equilibrium:** It signifies a dynamic equilibrium between the liquid (or solid) phase of water and its gaseous phase (water vapor). For every water molecule that evaporates, one condenses, resulting in no net change in the amount of water vapor in the air.
- **Relative Humidity of 100%:** When the air is saturated, the actual amount of water vapor present is equal to the maximum amount it can hold. This results in a relative humidity of 100%.
- **Onset of Condensation:** If saturated air is cooled further (without adding or removing water vapor), it will become supersaturated, leading to condensation (formation of liquid water like dew or fog) or deposition (formation of ice like frost). The temperature at which saturation occurs upon cooling is the dew point (for liquid water) or frost point (for ice).

In summary, saturation indicates that the air is „full" of water vapor for its current temperature, and any further addition of moisture or decrease in temperature will likely result in condensation. It's a critical threshold for the formation of clouds and precipitation.

8.3. Condensation and Sublimation

- **Condensation** is the process by which water vapor (gaseous state of water) in the air changes into liquid water (Fig. 8.1). This typically occurs when the air is cooled to its dew point or when the air becomes so saturated with water vapor that it can no longer hold it. The detailed explanation is:
    - **Cooling:** When air containing water vapor cools, the kinetic energy of the water molecules decreases. This allows the intermolecular forces between the water molecules to become more effective, causing them to come closer together and transition from a gas to a liquid state.
    - **Saturation:** Air has a certain capacity to hold water vapor, which is dependent on its temperature (warmer air can hold more). When the amount of water vapor in the air reaches this capacity (saturation),

any additional water vapor or a decrease in temperature will lead to condensation.

- **Condensation Nuclei:** In the atmosphere, condensation often occurs on tiny particles called condensation nuclei. These can be dust, pollen, salt crystals, or pollutants. Water vapor molecules are attracted to these surfaces, making it easier for them to coalesce into liquid droplets.

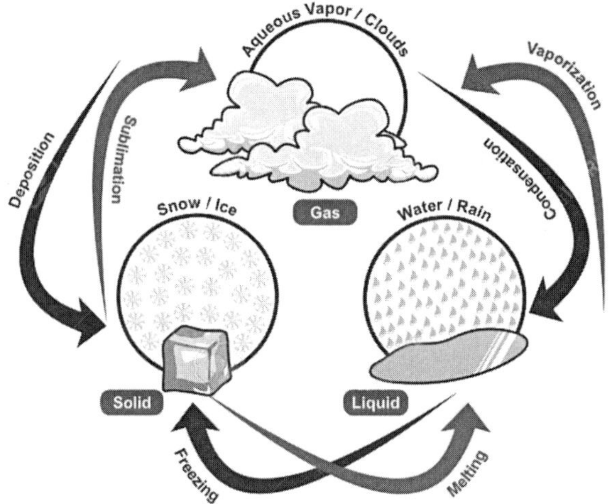

Fig. 8.1: Changes in states of water

- **Examples of Condensation in the Atmosphere**
  - **Dew:** Water droplets that form on cool surfaces overnight as the air near the surface cools to its dew point.
  - **Fog and Mist:** Clouds that form at ground level when the air near the surface cools and the water vapor condenses into tiny water droplets. Mist is like a light fog.
  - **Clouds:** Formed when moist air rises and cools in the atmosphere, causing water vapor to condense on condensation nuclei at higher altitudes.
- **Sublimation:** Sublimation is the process by which water changes directly from a solid state (ice or snow) to a gaseous state (water vapor) without passing through the liquid phase. The detailed explanation is:
  - **Energy Input:** Sublimation occurs when ice or snow absorbs enough energy (usually from sunlight or the surrounding air) for its molecules to overcome the intermolecular forces holding them in the solid structure and escape directly into the gaseous phase.

- **Bypassing the Liquid Phase:** Under certain conditions of temperature and pressure (typically below the triple point of water), the direct transition from solid to gas is energetically more favorable than melting followed by evaporation.

- **Examples of Sublimation in Nature**
  - **Shrinking Snow Piles:** On a sunny but cold day, snow can gradually disappear even if the temperature remains below freezing. This is because the ice is sublimating directly into water vapor.
  - **Drying Clothes in Freezing Temperatures:** Frozen clothes can dry over time because the ice in them sublimates.
  - **Formation of Some High-Altitude Clouds:** Ice crystals in very cold, high-altitude clouds can sometimes sublimate directly into water vapor if the surrounding air is dry enough.
  - **Frost Disappearing:** Similar to snow, frost (ice crystals formed on surfaces) can also sublimate.
  - In summary, **condensation** is the gas-to-liquid phase change of water, usually requiring cooling or saturation, often occurring on nuclei. **Sublimation** is the direct solid-to-gas phase change of water, requiring energy input and bypassing the liquid phase.

8.4. Freezing Point

In meteorology, the **freezing point** specifically refers to the temperature at which water transitions from a liquid state to a solid state (ice) at standard atmospheric pressure. This temperature is **0 degree Celsius (0°C) or 32 degrees Fahrenheit (32°F)**.

When we talk about the „freezing point“ in a meteorological context, it often relates to:

- **The Freezing Level:** This is the altitude in the atmosphere where the temperature is 0°C. Knowing the freezing level is crucial for determining the type of precipitation that will occur at the surface.
  - If the freezing level is high, precipitation falling from clouds that initially form as snow or ice may melt into rain before reaching the ground.
  - If the freezing level is low or at the surface, snow or ice is more likely.
  - If there‘s a layer of above-freezing air aloft and below-freezing air near the surface, freezing rain can occur (rain that freezes upon contact with surfaces below 0°C).

- **Surface Temperatures and Frost:** When surface temperatures drop to or below the freezing point, frost can form. Frost occurs through deposition, where water vapor in the air freezes directly onto surfaces that are below freezing.
- **Forecasting Winter Weather:** Predicting whether temperatures will fall below freezing is fundamental to forecasting snow, sleet, freezing rain, and ice storms.
- **Aviation:** The freezing level is important for aviation as it indicates where icing conditions might be present.Aircraft icing can be hazardous.

In essence, "freezing point" in meteorology is the critical temperature of 0°C that dictates whether water will exist as a liquid or solid, influencing precipitation type, surface conditions, and atmospheric hazards.

8.5. Relative Humidity

Relative humidity (RH) is a measure of the amount of water vapor present in the air compared to the maximum amount of water vapor the air could hold at a given temperature. It is expressed as a percentage (Fig. 8.2).

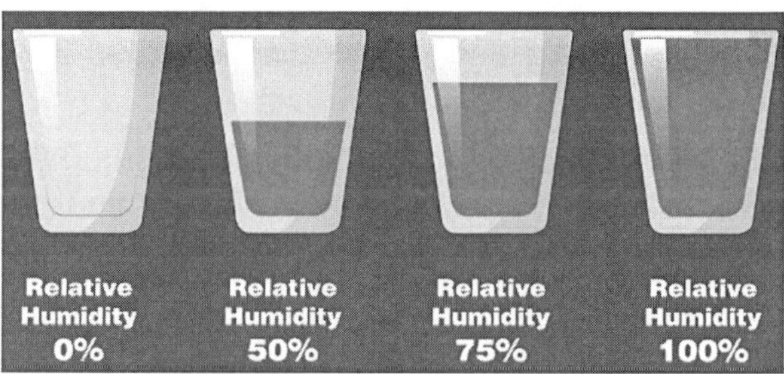

Fig. 8.2: Relative humidity

Think of it this way: warmer air has a greater capacity to hold water vapor than cooler air. Relative humidity tells you how „full" the air is with water vapor relative to its total capacity at that specific temperature.

- **100 % RH** means the air is saturated and cannot hold any more water vapor. At this point, condensation (or deposition) is likely to occur, leading to the formation of dew, fog, or clouds.
- **0 % RH** means the air is completely dry and contains no water vapor.

8.5.1. Formula to Calculate Relative Humidity

Relative humidity can be calculated in a few ways, but the most common conceptual formula is:

RH = (Actual amount of water vapor in the air / Maximum amount of water vapor the air can hold at that temperature) * 100 %

More formally, using vapor pressures

RH = (Actual Vapor Pressure (e) / Saturated Vapor Pressure (es)) * 100 %

Where:

- (e) is the partial pressure exerted by the water vapor in the air.
- (e_s) is the maximum vapor pressure that water can exert at that specific temperature (saturated vapor pressure).

There are also more complex formulas that use temperature and dew point to calculate relative humidity, as seen in some of the search results provided, but the core concept remains the ratio of the actual moisture to the potential moisture.

Based on the current weather in Delhi, the humidity is 55%.

8.5.2. Significance of Relative Humidity

Relative humidity is a crucial atmospheric variable for several reasons:

- **Human Comfort:** It significantly affects how we perceive temperature. High relative humidity makes us feel hotter because the evaporation of sweat from our skin is reduced. Conversely, low relative humidity can make us feel cooler due to increased evaporation. The „feels like" temperature often incorporates the effect of humidity.
- **Weather Forecasting:** RH is a key indicator of the likelihood of precipitation, dew, and fog. High RH suggests that the air is close to saturation, making condensation more likely with further cooling or the addition of more moisture.
- **Agriculture:** It influences plant transpiration rates and the likelihood of dew formation, which can impact plant health and the spread of diseases.
- **Material Science and Preservation:** RH affects the moisture content of various materials, influencing their properties, stability, and longevity. For example, high RH can lead to corrosion, mold growth, and the swelling of wood. Museums and storage facilities often carefully control RH to preserve artifacts.
- **Health:** Both very high and very low relative humidity can impact human health. High humidity can exacerbate respiratory issues, while low humidity can dry out skin and mucous membranes, increasing susceptibility to infections.10

In summary, relative humidity provides a measure of how close the air is to being saturated with water vapor, which has significant implications for our comfort, the weather, and various natural and human-made systems.

8.6. Specific Humidity

Specific humidity is a measure of the mass of water vapor in a given mass of air, including the water vapor itself. It's an important concept in meteorology and climatology. The formula is:

Specific humidity (q) = mass of water vapor / total mass of air (including water vapor)

$q = mv / (md + mv)$

where: mv = mass of water vapor; md = mass of dry air

- **Meaning:** Specific humidity represents the actual amount of water vapor present in a unit mass of air. It's usually expressed in grams per kilogram (g/kg).
- **Significance**
  - **Weather forecasting:** Specific humidity helps predict precipitation, fog, and other weather phenomena.
  - **Climate modeling:** It's used to understand and model global climate patterns, including the water cycle and atmospheric circulation.
  - **Agriculture:** Specific humidity affects crop growth, evapotrans piration, and irrigation needs.
  - **Aviation:** It's crucial for predicting weather conditions that impact flight operations.

In summary, specific humidity is a key parameter in understanding atmospheric conditions and predicting weather patterns.

9

Cloud Formation and Classification

Clouds, those ever-present features of our sky, are more than just aesthetically pleasing; they are fundamental components of earth's weather and climate systems. This chapter, "cloud formation and classification," will embark on a journey to understand the fascinating processes that lead to their creation and the systematic ways in which they are categorized.

We will begin by exploring the essential ingredients for cloud formation: the presence of water vapor in the atmosphere, the availability of condensation or ice nuclei, and the crucial mechanism of air cooling. We will delve into how rising air parcels cool adiabatically, reaching saturation and allowing water vapor to condense or deposit into the myriad forms we recognize as clouds.

Following the understanding of their genesis, we will navigate the internationally recognized system of cloud classification. This system, primarily based on a cloud's altitude, appearance, and the processes that formed it, provides a common language for meteorologists and observers worldwide. We will learn to identify the main cloud types – high, mid-level, low, and vertically developing – and their associated characteristics, from the wispy cirrus to the towering cumulonimbus.

Furthermore, we will touch upon the significance of different cloud types in indicating current weather conditions and forecasting future weather. The appearance and evolution of clouds can offer valuable clues about atmospheric stability, moisture content, and the likelihood of precipitation.

By the end of this chapter, you will gain a comprehensive understanding of how clouds form the diverse array of cloud types that grace our skies, and their importance in the broader context of atmospheric science. Let's unlock the secrets held within these visible masses of water and ice suspended in our atmosphere.

9.1. Clouds

Clouds are visible masses of tiny water droplets or ice crystals suspended in the Earth's atmosphere. These droplets or crystals are so small and light that they can remain aloft.What we see as a cloud is essentially a large collection of these minute particles.

9.1.1. Formation of Clouds

Clouds form when invisible water vapor in the air condenses into liquid water droplets or freezes into ice crystals. This happens when air becomes saturated, meaning it can no longer hold all of its water vapor (Fig. 9.1). Saturation is typically reached in one of two ways:

- **Cooling the Air:** As air rises in the atmosphere, it expands due to lower air pressure at higher altitudes. This expansion causes the air to cool. Cooler air can hold less water vapor than warmer air. When the air cools to its **dew point** (the temperature at which the air becomes saturated), the excess water vapor condenses.
- **Adding Moisture to the Air:** If more water vapor is added to the air (e.g., through evaporation from a body of water), the air can eventually become saturated at its current temperature, leading to condensation.

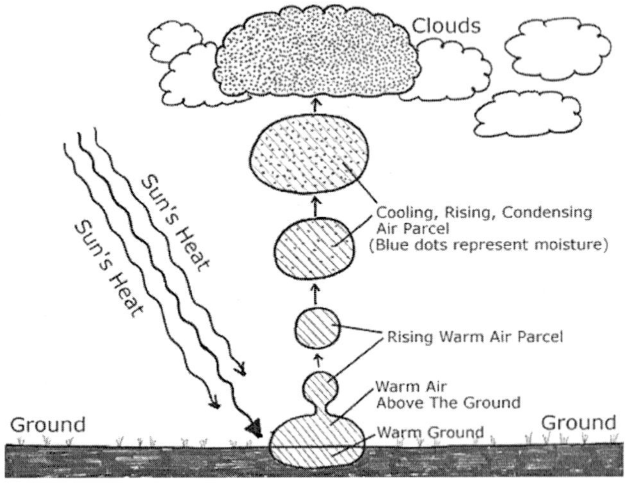

Fig. 9.1: Formation of clouds

For condensation (or deposition, the direct formation of ice from vapor) to occur, water vapor needs a surface to condense onto. These microscopic airborne particles, such as dust, pollen, salt crystals, and pollutants, are called **condensation nuclei** (or ice nuclei if the temperature is below freezing). Water vapor molecules cluster around these nuclei to form the tiny cloud droplets or ice crystals.

9.1.2. Growth of Clouds

Once the initial cloud droplets or ice crystals form, clouds can grow through several processes:

- **Further Condensation/Deposition:** If the air continues to cool or more moisture is added, more water vapor will condense or deposit onto the existing droplets or crystals, causing them to grow larger.
- **Collision and Coalescence:** Within a cloud, droplets of different sizes fall at slightly different speeds. This can lead to collisions between droplets. When they collide, they can coalesce (merge) into larger droplets. This process is particularly important in warmer clouds for the formation of raindrops.
- **Bergeron-Findeisen Process (Ice-Crystal Process):** In colder clouds (where both ice crystals and supercooled water droplets coexist), ice crystals tend to grow at the expense of water droplets. This is because the saturation vapor pressure over ice is lower than that over liquid water at the same temperature. Water vapor will deposit onto the ice crystals more readily than it will condense onto the water droplets, causing the ice crystals to grow. These ice crystals can then fall and melt into rain, or remain as snow or other forms of frozen precipitation.
- **Uplift:** Continued lifting of air can lead to more condensation as the air cools at higher altitudes, adding more moisture to the cloud and increasing its vertical extent. This is particularly important for the development of towering clouds like cumulonimbus.

The growth of clouds is a dynamic process involving the interplay of thermodynamics, microphysics (the behavior of tiny water and ice particles), and atmospheric motion.

9.1.3. Dissipation of Clouds

It refers to the process by which clouds weaken, shrink, and eventually disappear from the sky. This happens when the conditions that led to their formation and growth cease to be effective, and the balance shifts towards the evaporation or sublimation of the water droplets or ice crystals within them. Here are the main ways clouds dissipate:

- **Warming of the Air:** If the temperature of the air within or around the cloud increases, the air's capacity to hold water vapor rises. This can lead to the evaporation of the water droplets or sublimation of the ice crystals in the cloud, causing it to shrink and disappear. Solar heating, especially for low clouds, or the descent of air (which warms adiabatically) can cause this.
- **Mixing with Drier Air (Entrainment):** Clouds don't exist in isolation; they constantly interact with the surrounding air. If the air surrounding a cloud is significantly drier (lower relative humidity), the water droplets in the cloud will evaporate into this drier air in an attempt to reach

equilibrium. This mixing of dry air into the cloud, known as entrainment, can erode the cloud over time, making its edges appear wispy and less defined until it vanishes.

- **Sinking Air (Subsidence):** When air sinks in the atmosphere, it compresses and warms adiabatically. This warming increases the air's capacity to hold water vapor, leading to the evaporation of cloud droplets. Large-scale sinking motion associated with high-pressure systems can cause widespread cloud dissipation.
- **Precipitation:** If a cloud produces rain, snow, or other forms of precipitation, it loses some of its water content. If the rate of precipitation exceeds the rate at which water vapor is condensing within the cloud, the cloud will eventually dissipate as its water supply is depleted. Essentially, the cloud "rains itself out."

Often, the dissipation of a cloud involves a combination of these processes working simultaneously. For example, a low cloud might dissipate due to a combination of daytime heating and mixing with slightly drier air above it.

9.1.4. Classification of Clouds

The primary criteria used for the classification of clouds are their **altitude of formation** in the atmosphere and their **appearance or physical form**. This system, largely based on the work of Luke Howard in the early 19th century and refined by the World Meteorological Organization (WMO), allows for a systematic understanding and identification of different cloud types (Fig. 9.2). Here's a breakdown of the criteria:

- **Altitude of Formation:** Clouds are broadly categorized into three main height levels (though some extend through multiple levels):
    - **High-Level Clouds:** Typically form above 5,000 meters (16,500 feet) in temperate regions. At these altitudes, the air is very cold, so these clouds are primarily composed of ice crystals. They are given the prefix "cirro-". Examples include Cirrus (Ci), Cirrocumulus (Cc), and Cirrostratus (Cs).
    - **Mid-Level Clouds:** Form between 2,000 and 7,000 meters (6,500 to 23,000 feet) in temperate regions. These clouds are composed of water droplets, ice crystals, or a mixture of both (often supercooled water). They are given the prefix "alto-". Examples include Altocumulus (Ac) and Altostratus (As).
    - **Low-Level Clouds:** Form from the surface up to 2,000 meters (6,500 feet). These are mainly composed of water droplets, but can contain ice crystals in colder regions or during winter. Examples include

Stratus (St), Stratocumulus (Sc), and Nimbostratus (Ns - which also has vertical development).

- **Clouds with Vertical Development:** These clouds don't fit neatly into a single height category as their bases are typically in the low to mid-levels, but their tops can extend high into the atmosphere. They are associated with instability in the atmosphere. Examples include Cumulus (Cu) and Cumulonimbus (Cb).

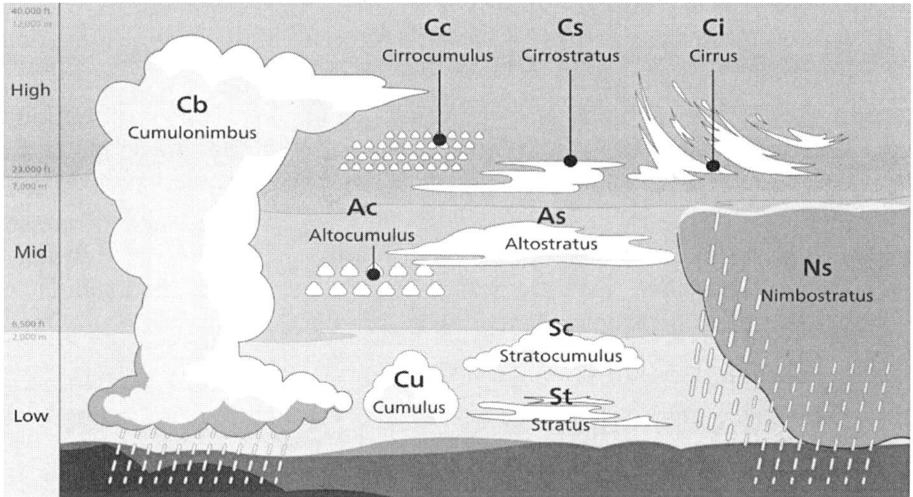

Fig. 9.2: Clouds classification

The altitude ranges can vary slightly depending on latitude (they tend to be higher in the tropics and lower in Polar Regions).

- **Appearance or Physical Form:** The second key criterion is the visual characteristics of the clouds, which describe their shape, texture, and opacity. Several root names are used to describe these forms:
  - **Cirrus (Ci):** Thin, wispy, and feathery clouds composed of ice crystals. The term "cirro-" is used for high-level clouds of this nature.
  - **Cumulus (Cu):** Puffy, cotton-like clouds with flat bases and dome-shaped tops, indicating rising air. The term "cumulo-" is used in combination with other height indicators.
  - **Stratus (St):** Flat, hazy, and featureless sheet-like clouds that often cover the entire sky. The term "strato-" is used for low-level clouds of this nature.
  - **Nimbus (Nimbo-):** Indicates a cloud that is producing or capable of producing precipitation (rain or snow). This prefix or suffix is added to other cloud type names (e.g., Nimbostratus, Cumulonimbus).

By combining these altitude-based prefixes/names with the form-based names, we get the names of the ten basic cloud genera recognized internationally. For example:

- **Cirrostratus:** High-level (cirro-) and sheet-like (stratus).
- **Altocumulus:** Mid-level (alto-) and puffy (cumulus).
- **Nimbostratus:** Low to mid-level (stratus-based) and rain-bearing (nimbo-).
- **Cumulonimbus:** Vertically developing (cumulo-) and rain-bearing (nimbus).

Beyond these genera, clouds are further classified into **species** (describing more specific shapes and internal structures) and **varieties** (describing the arrangement and transparency). Additional features like precipitation (virga, praecipitatio), and accessory clouds (pileus, mamma) are also noted.

In summary, the classification of clouds primarily relies on **how high they are** and **what they look like**, providing a structured way to understand and communicate about these important atmospheric phenomena.

9.2. Description of Main Genera of Clouds

Let us rewrite the descriptions for the main cloud genera, incorporating both their altitude and appearance:

- **High-Level Clouds (composed primarily of ice crystals)**
    - **Cirrus (Ci):** These are delicate, white, often fibrous or silky clouds with a wispy appearance. They often look like streaks or thin, detached patches with hair-like filaments. Being high up, they are made of ice crystals and typically appear against a blue sky. They can indicate an approaching warm front.
    - **Cirrocumulus (Cc):** Appearing high in the atmosphere, these clouds look like small, white patches arranged in ripples or waves, sometimes described as a "mackerel sky." They are composed of ice crystals and can sometimes show iridescence (colors).
    - **Cirrostratus (Cs):** These are thin, sheet-like, high-level clouds composed of ice crystals. They often appear as a pale, milky veil that can cover the entire sky. A common characteristic is the halo effect they can produce around the sun or moon due to the refraction of light through their ice crystals.
- **Mid-Level Clouds (composed of water droplets, ice crystals, or a mixture)**
    - **Altocumulus (Ac):** These mid-level clouds appear as white or gray patches, often arranged in layers or waves. They can have a puffy

or somewhat fibrous appearance. Sometimes, parts of the cloud may be darker than others. They are often described as being larger and thicker than cirrocumulus.

- **Altostratus (As):** These are gray or bluish-gray mid-level sheets that often cover the entire sky. The sun or moon may be visible as a dim, watery disk without a halo. Altostratus clouds are typically more uniform and less structured than altocumulus. They can indicate an approaching storm system.
- **Low-Level Clouds (composed primarily of water droplets)**
  - **Stratus (St):** These are gray, uniform, sheet-like low-level clouds that can often cover the entire sky, resembling a featureless overcast. They may produce light drizzle or mist. When stratus clouds are very low and in contact with the ground, they are called fog.
  - **Stratocumulus (Sc):** These low-level clouds appear as patches or layers of rounded masses, rolls, or ripples. They are generally gray or whitish with darker parts. Stratocumulus clouds are more defined and have more texture than stratus clouds. They rarely produce significant precipitation.
  - **Nimbostratus (Ns):** These are dark, gray, featureless rain clouds (or snow clouds). They are typically thick enough to block out the sun entirely and are associated with continuous, light to moderate precipitation that can last for several hours. While their bases are low, they can also extend into the mid-levels.
- **Clouds with Vertical Development (bases in low to mid-levels, tops can extend high):**
  - **Cumulus (Cu):** These are puffy, dome-shaped clouds with flat bases. They are often described as looking like cotton balls. Fair-weather cumulus clouds (Cumulus humilis) are small and don't typically produce precipitation. However, they can grow vertically into larger Cumulus congestus, which can bring showers.
  - **Cumulonimbus (Cb):** These are large, towering, vertically developed clouds associated with thunderstorms. Their tops can reach very high into the atmosphere, often taking on an anvil shape (incus). Cumulonimbus clouds are capable of producing heavy rain, lightning, thunder, hail, and even tornadoes. Their bases are often dark and menacing.

10

Processes of Condensation

Let's dive deeper into the processes of condensation and explore each in more detail:

- **Adiabatic Processes:** Adiabatic cooling occurs when air rises and expands, reducing its temperature (Fig. 10.1). This process can lead to condensation and cloud formation. There are several ways air can rise and cool adiabatically:

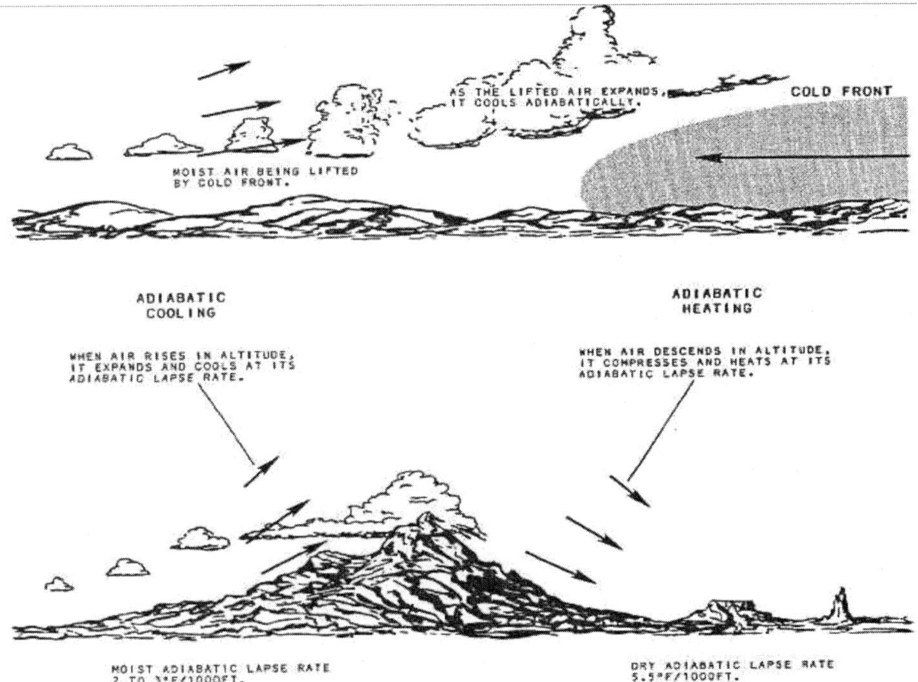

Fig. 10.1: Adiabatic processes

- **Orographic Lift:** When air is forced to rise over a mountain or hill, it cools, and condensation occurs, resulting in clouds and precipitation. This is why mountainous regions often experience more rainfall than surrounding are as.

  - **Frontal Lift:** When two air masses with different temperatures and humidity levels meet, the warmer air is forced to rise, cool, and condense, forming clouds and precipitation. This can lead to significant weather changes, such as the formation of fronts and low-pressure systems.
  - **Convection:** When the sun heats the ground, it warms the air closest to the surface, causing it to rise and cool, leading to condensation and cloud formation. This process can lead to the development of cumulus clouds, thunderstorms, and other types of convective weather.
- **Non-Adiabatic Processes:** Non-adiabatic cooling occurs when air cools without rising. There are several ways this can happen:
  - **Radiation Cooling:** When the ground rapidly loses heat by radiation, the air above it cools, and condensation occurs, forming dew, frost, or fog. This process typically occurs overnight, especially in clear skies and calm conditions.
  - **Conduction:** When air comes into contact with a cold surface, it cools, and condensation occurs, forming dew, frost, or fog. This can happen when warm air moves over a cold surface, such as a lake or ocean.
  - **Advection:** When warm air moves over a cold surface, it cools, and condensation occurs, forming fog or clouds. This can happen when a warm front moves over a cold region or when a cold ocean current meets warm air.

10.1. Forms of Condensation

- **Dew:** Dew forms when moisture in the air condenses on surfaces as the temperature cools overnight. This process requires clear skies, calm air, and high relative humidity. Dew can be an important source of moisture for plants and animals (Fig. 10.2).

Fig. 10.2: Dew deposition on plant leaf

- **Frost:** Frost forms when condensation occurs below freezing point, resulting in minute ice crystals (Fig. 10.3). This can damage crops and plants, especially if the frost is severe or prolonged.

Fig. 10.3: Frost formation on grass

- **Fog:** Fog forms when water vapor condenses on fine dust particles or other nuclei in the air, reducing visibility. There are several types of fog, including:
  - **Radiation Fog:** Forms overnight through radiative cooling, typically in valleys and low-lying areas. This type of fog can be dense and persistent.
  - **Advection Fog:** Forms when warm air moves over a cold surface, such as a cold ocean current. This type of fog can be widespread and persistent.
  - **Frontal or Precipitation Fog:** Forms when warm and cold air masses converge, resulting in precipitation and fog. This type of fog can be associated with significant weather changes.
  - **Steam Fog:** Forms when cold air moves over warm water, resulting in rapid evaporation and condensation. This type of fog can be dense and localized.
  - **Ice Fog:** Forms when water vapor freezes into tiny ice crystals in cold air. This type of fog can be associated with very low temperatures.
- **Clouds:** Clouds form when water vapor condenses in free air at considerable elevations. Clouds can be classified into various types based on height, shape, and other characteristics, such as:
  - **Cirrus Clouds:** High-level clouds composed of ice crystals. These clouds can indicate fair weather or approaching storms.
  - **Cumulus Clouds:** Puffy, white clouds that can grow into towering cumulus or thunderstorms. These clouds can indicate fair weather or instability in the atmosphere.

- **Stratus Clouds:** Low-level clouds that often cover the entire sky and produce light to moderate precipitation. These clouds can indicate overcast weather or approaching storms.
- **Nimbus Clouds:** Dark, rain-bearing clouds that can produce heavy precipitation. These clouds can indicate significant weather changes.

10.2. Factors Influencing Condensation

- **Temperature:** Warm air can hold more water vapor than cold air. When air cools, its capacity to hold water vapor decreases, leading to condensation.
- **Humidity:** High humidity means the air is close to saturation, making condensation more likely. Low humidity means the air can hold more water vapor, making condensation less likely.
- **Surface Area:** Increased surface area can facilitate condensation, as seen in the formation of dew on grass or leaves.
- **Air Movement:** Calm conditions can lead to the formation of dew or frost, while windy conditions can disrupt the condensation process.
- **Atmospheric Pressure:** Changes in atmospheric pressure can influence the formation of clouds and precipitation.

10.3. Importance of Condensation

- **Water Cycle:** Condensation plays a crucial role in the Earth's water cycle, as it helps distribute water from the atmosphere to the land and oceans.
- **Weather Patterns:** Condensation influences weather patterns, including precipitation, fog, and clouds.
- **Agriculture:** Condensation affects crop growth, as dew and precipitation provide plants with water.
- **Climate Regulation:** Condensation helps regulate Earth's climate by influencing the amount of solar radiation that reaches the surface.
- **Ecosystems:** Condensation supports ecosystems, as it provides water for plants and animals.

In conclusion, condensation is a complex and multifaceted process that plays a crucial role in shaping our climate, weather, and ecosystems. Understanding condensation is essential for predicting weather patterns, managing water resources, and mitigating the impacts of climate change.

11

Precipitation and Its Forms

Precipitation is a vital component of the Earth's water cycle, playing a crucial role in shaping our climate, weather, and ecosystems. It occurs when water vapor in the atmosphere condenses and falls to the ground as rain, snow, sleet, or hail. Precipitation is a key driver of weather patterns, influencing everything from local weather events to global climate trends.

In this chapter, we will explore the various forms of precipitation, including rain, snow, sleet, and hail, and examine the factors that influence their formation and distribution (Fig. 11.1).

Fig. 11.1: Forms of precipitation

We will also discuss the importance of precipitation in sustaining life on Earth, from supporting plant growth and agriculture to shaping landscapes and ecosystems. Key topics to be discussed are:

- Forms of precipitation (rain, snow, sleet, hail)
- Precipitation processes (formation, distribution, intensity)
- Factors influencing precipitation (temperature, humidity, wind patterns)
- Importance of precipitation (ecosystems, agriculture, water resources)
- Precipitation measurement and prediction (methods, challenges, applications)

By understanding precipitation, we can better appreciate the complex interactions between the atmosphere, oceans, and land surfaces that shape our planet's climate and weather patterns.

11.1. Forms of Precipitation

11.1.1. Rain

Rain is a type of precipitation that occurs when water droplets in clouds grow and become too heavy to remain suspended in the air. Rain droplets are typically larger than 0.5 mm in diameter. Rainfall can be classified into different types, including convectional, orographic, and cyclonic rainfall.

11.1.2. Drizzle

Drizzle is a type of light rainfall with drop sizes less than 0.5 mm. It's often steady and continuous, but with much smaller droplets than regular rain.

11.1.3. Snow

Snow forms when water vapor in clouds freezes into ice crystals. Snowfall occurs when the temperature is below 0°C, and the ice crystals stick together to form snowflakes. Snow can be heavy or light and its intensity can vary greatly.

11.1.4 Sleet

Sleet forms when snowflakes fall through a layer of warm air, causing them to melt into raindrops. If these raindrops then pass through a layer of cold air before reaching the ground, they freeze into small, transparent ice pellets called sleet.

11.1.5. Hail

Hail forms when updrafts in thunderstorms carry water droplets up into the freezing level of the atmosphere, where they freeze into small balls of ice. As the hailstones move upward and downward through the storm cloud, they may pass through additional layers of water droplets and freezing air, causing them to grow in size.

11.1.6. Graupel

A small, white ice particle that falls as precipitation and breaks apart easily when it lands on a surface. Also called snow pellet soft hail.

11.1.7. Frost

Frost forms when the air temperature cools to its dew point, causing the water vapor to condense and freeze onto surfaces. This can occur when the air temperature is below freezing, resulting in a layer of ice crystals on surfaces.

11.1.8. Dew

Dew forms when the air temperature cools to its dew point, causing the water vapor to condense onto surfaces. This typically occurs overnight as the ground rapidly cools, causing the air near the surface to cool and reach its dew point.

11.1.9. Other Forms of Precipitation

- **Virage**: Raindrops that evaporate before reaching the ground, often occurring when rain falls through dry air.
- **Mist**: A type of fog that forms when evaporation occurs before reaching the ground, reducing visibility.
- **Hailstones**: Large balls of ice that form in thunderstorms and can cause significant damage.

Precipitation forms through various processes (Fig. 11.2), including:

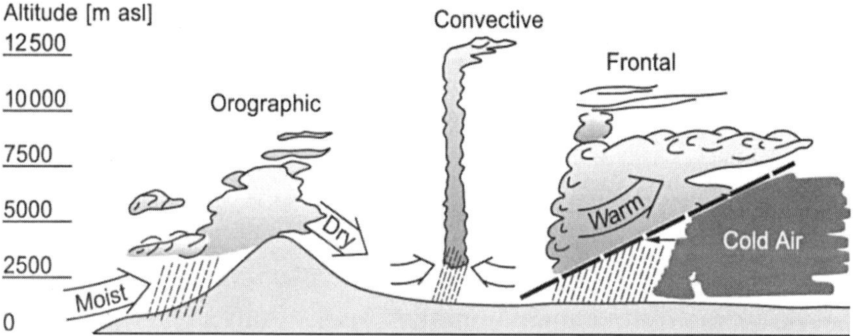

Fig. 11.2: Precipitation through various processes

- **Convectional Precipitation**: Heated air rises, cools, and condenses, forming clouds and precipitation.
- **Orographic Precipitation**: Moist air is forced to rise over mountains, cooling and condensing to form clouds and precipitation.
- **Cyclonic Precipitation**: Large-scale precipitation that occurs when air masses with different temperatures and humidity levels meet like frontal masses.

11.2. Convectional Precipitation

Convectional precipitation is a type of precipitation that occurs when the sun's energy heats the surface of the Earth, causing water to evaporate and form water vapor. This process leads to the formation of clouds and precipitation through a series of stages. The Process of convectional precipitation is:

- **Evaporation:** The sun heats the ground, causing water to evaporate from surfaces such as oceans, lakes, and rivers. This water vapor rises into the air as gas.
- **Rising Air:** As the ground continues to heat up, it warms the air above it, causing it to expand and rise in convection currents. This warm air is less dense than the surrounding air, allowing it to rise.
- **Cooling and Condensation:** As the air rises, it cools, and the water vapor in the air condenses into tiny water droplets, forming clouds.
- **Cloud Formation:** The condensed water droplets gather together to form visible clouds. Cumulonimbus clouds are a common type of cloud associated with convectional precipitation.
- **Precipitation:** When the water droplets in the clouds become too heavy to remain suspended in the air, they fall to the ground as precipitation, often in the form of heavy rain or thunderstorms.

11.2.1. Characteristics of Convectional Precipitation

- **Heavy Rainfall:** Convectional precipitation is characterized by heavy rainfall, often accompanied by thunder and lightning.
- **Short-Lived:** This type of precipitation is typically short-lived, lasting from a few minutes to a few hours.
- **Localized:** Convectional precipitation is highly localized, often occurring in specific areas where the ground is heated by the sun.
- **Associated with Cumulonimbus Clouds:** Convectional precipitation is often associated with cumulonimbus clouds, which can produce heavy precipitation and thunderstorms.

11.2.2. Regions Prone to Convectional Precipitation

- **Tropical Regions:** Convectional precipitation is common in tropical regions near the equator, where the sun's energy is strongest.
- **Equatorial Regions:** Regions like the Amazon Rainforest, Congo Basin, and islands of Southeast Asia experience convectional precipitation due to their location near the equator.
- **Temperate Regions During Summer:** Areas like South East England can experience convectional precipitation during warm sunny spells in the summer.

11.2.3. Importance of Convectional Precipitation

- **Water Cycle:** Convectional precipitation plays a crucial role in the water cycle, helping to distribute water around the globe.

- **Agriculture:** This type of precipitation can provide much-needed water for crops, especially in regions with limited rainfall.
- **Weather Patterns:** Convectional precipitation can influence local weather patterns, leading to the formation of thunderstorms and heavy rainfall.

In conclusion, convectional precipitation is an important process that helps shape our climate and weather patterns. Understanding this process can provide valuable insights into the complex interactions between the atmosphere, oceans, and land surfaces.

11.3. Orographic Precipitation

Orographic precipitation occurs when a saturated air mass encounters a mountain or orographic barrier, forcing it to rise and cool, resulting in condensation and precipitation. This type of precipitation is commonly observed in mountainous regions where moist air is forced to ascend, leading to cloud formation and precipitation. The process of orographic precipitation is:

- **Air Mass Encounter**: A saturated air mass encounters a mountain or orographic barrier, such as a mountain range.
- **Forced Ascent**: The air mass is forced to rise due to the initial momentum, leading to a decrease in pressure and temperature.
- **Condensation**: As the air mass gains height, the water vapor condenses into clouds, and precipitation occurs when the clouds become saturated.
- **Precipitation Distribution**: The windward slopes of the mountain receive more precipitation, while the leeward slopes remain dry due to the descending air.

11.3.1. Characteristics of Orographic Precipitation

- **Windward Slopes Receive More Rainfall**: Mountains and hills can force warm, moist air to rise, cool, and condense, resulting in precipitation on the windward side.
- **Rain Shadow Effect**: The leeward side of the mountain receives less rainfall due to the descending air, creating arid or semi-arid regions.
- **Localized Precipitation**: Orographic precipitation is highly localized, often occurring in specific areas where the terrain forces air to rise.

11.3.2. Examples of Orographic Precipitation

- **Mahabaleshwar, India**: Situated on the Western Ghats, receives over 600 cm of rainfall annually, while Pune, in the rain shadow area, receives only about 70 cm.

- **Patagonian Desert, Argentina**: An example of a region affected by the rain shadow effect, resulting in arid conditions.

11.3.3. Importance of Orographic Precipitation

- **Water Cycle**: Orographic precipitation plays a crucial role in the water cycle, distributing water around the globe.
- **Ecosystems**: This type of precipitation supports plant growth and agriculture in regions with limited rainfall.
- **Climate Regulation**: Orographic precipitation influences local climate patterns, shaping regional weather conditions .

11.4. Cyclonic Precipitation

The process of cyclonic precipitation involves several steps:

- **Low Pressure Formation**: A low-pressure system develops due to the warming of air near the surface or the convergence of air masses.
- **Air Rise**: As air rises in the low-pressure system, it moves into cooler parts of the atmosphere.
- **Condensation**: The rising air cools, and the water vapor condenses into clouds.
- **Precipitation**: When the clouds are saturated with water, precipitation occurs.

11.4.1. Characteristics

- **Intensity and Duration**: Cyclonic precipitation can vary widely in intensity and duration. Tropical cyclones are known for intense, short-duration rainfall, while temperate cyclones can bring prolonged periods of rain or snow.
- **Distribution**: The precipitation is often widespread around the center of the cyclone but can be more concentrated near the eye wall in tropical cyclones.
- **Impact**: Cyclonic precipitation can lead to significant flooding, damage from strong winds, and disruptions in transportation and daily life.

11.4.2. Examples and Impacts

Cyclonic precipitation is a major cause of natural disasters worldwide. For instance, hurricanes in the Atlantic and Indian Ocean cyclones have been responsible for devastating floods and loss of life in coastal regions. Understanding and predicting cyclonic precipitation patterns are crucial for mitigating these impacts through early warning systems and preparedness measures.

11.5. Artificial Rain Making

Artificial rain making, also known as cloud seeding or weather modification, is the technique of artificially inducing or increasing precipitation from clouds.1 This is achieved by dispersing substances into the air that serve as cloud condensation or ice nuclei, thus altering the microphysical processes within the cloud.

11.5.1. How It Is Done?

The primary method of artificial rain making is **cloud seeding**. This involves introducing agents into existing clouds to stimulate precipitation (Fig. 11.3).

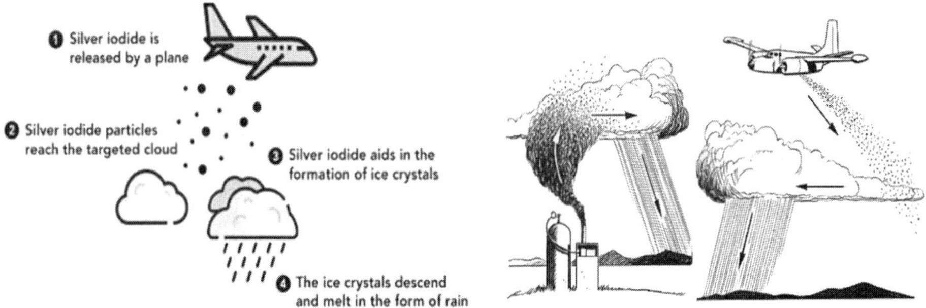

Fig. 11.3: Artificial rain making

The common techniques include

- **Static Cloud Seeding (Glaciogenic Seeding)**
  - This method targets supercooled clouds (clouds with temperatures below freezing but containing liquid water).
  - Agents like **silver iodide (AgI)** or **dry ice (solid carbon dioxide)** are dispersed into the cloud.
  - Silver iodide particles have a crystalline structure similar to ice, acting as ice nuclei around which water vapor can freeze, forming ice crystals.
  - Dry ice cools the air significantly, causing water vapor to freeze spontaneously into ice crystals. These ice crystals then grow by accretion of supercooled water droplets. If the lower atmosphere is warm enough, these ice crystals melt as they fall, resulting in rain.
  - Dispersion can be done via aircraft that fly through or above the cloud, releasing the seeding agent through flares or dispensers, or from ground-based generators that release silver iodide smoke carried upward by air currents.

- **Hygroscopic Cloud Seeding (Warm Cloud Seeding)**
  - This method targets warmer clouds that are composed entirely of liquid water.
  - Hygroscopic materials like **salts (e.g., sodium chloride, calcium chloride)** are introduced into the lower parts of the clouds.
  - These salt particles attract water vapor, causing the cloud droplets to grow in size through coalescence (merging with other droplets). When they become heavy enough, they fall as rain.
  - Dispersion is typically done using flares or explosives from aircraft or ground-based methods.
- **Dynamic Cloud Seeding**
  - This is a more complex technique aiming to enhance the convective activity within clouds, encouraging more cloud growth and precipitation.
  - It often involves seeding the upper parts of clouds with larger quantities of ice nuclei (compared to static seeding) to promote the Bergeron-Findeisen process and invigorate the cloud's dynamics.
  - Newer approaches also include using drones to deliver electric charges to stimulate rainfall or infrared laser pulses aimed at inducing particle formation.
- **Conditions Required:** The success of artificial rain making largely depends on the presence of suitable atmospheric conditions and cloud characteristics:
  - **Presence of Clouds:** There must be existing clouds with sufficient moisture content. Cloud seeding cannot create clouds.
  - **Supercooled Water (for Glaciogenic Seeding):** For silver iodide or dry ice seeding to be effective, the clouds should ideally contain supercooled water droplets (water in liquid form at temperatures below freezing).
  - **Sufficient Updrafts (for Dynamic Seeding):** For dynamic seeding to work, there should be sufficient vertical air currents within the clouds to allow for enhanced growth and precipitation.
  - **Warm Cloud Characteristics (for Hygroscopic Seeding):** For salt seeding, warmer clouds with a high liquid water content are preferred.
  - **Favorable Wind Patterns:** The wind should be such that the effects of seeding are observed in the intended target area.

- **Appropriate Temperature Profiles:** The atmospheric temperature profile below the cloud base needs to be conducive to the precipitation reaching the ground as rain (i.e., not freezing all the way down if ice crystals are formed).

- **Advantages:** Artificial rain making offers several potential benefits:
  - **Drought Relief:** In regions facing prolonged dry spells and water scarcity, cloud seeding can augment natural rainfall, providing much-needed water for agriculture, domestic use, and replenishing water bodies like reservoirs and lakes.
  - This can help mitigate the severe impacts of drought on ecosystems and human populations, reducing crop failures and water shortages.
  - **Increased Water Supply:** By enhancing precipitation, especially in mountainous regions, cloud seeding can contribute to increased snowpack during winter.This snowpack melts in the spring and summer, providing a crucial source of water for downstream areas.
  - This increased runoff can benefit hydroelectric power generation by ensuring higher water levels in reservoirs.
  - **Agricultural Benefits:** Timely artificial rainfall can support agricultural productivity by providing essential water for crops, especially during critical growth stages or when natural rainfall is insufficient. This can lead to improved crop yields and reduced dependence on irrigation in some areas.
- **Hail Suppression:** In areas prone to damaging hailstorms, cloud seeding techniques can be employed to reduce the size of hailstones.By introducing a large number of ice nuclei, the supercooled water in the storm clouds freezes onto these nuclei, resulting in a greater number of smaller hailstones, which are less damaging than fewer, larger ones.
- **Fog Dispersal:** Cloud seeding can be used to clear fog, particularly at airports and in other areas where low visibility poses a safety hazard. Seeding with hygroscopic materials can cause the fog droplets to coalesce and precipitate out.
- **Air Pollution Control (Potentially):** In urban areas suffering from severe air pollution, artificial rain has been explored as a short-term solution to wash pollutants out of the atmosphere.The falling raindrops can capture particulate matter and gases, temporarily improving air quality.mHowever, this is not a long-term solution and addresses the symptom rather than the cause.

- **Wildfire Management:** In extremely dry conditions, artificial rain could potentially dampen vegetation and reduce the risk or intensity of wildfires, providing valuable time for firefighting efforts.

In summary, artificial rain making, primarily through cloud seeding, is a weather modification technique with the potential to alleviate water scarcity, support agriculture, mitigate weather hazards, and even address air pollution to some extent. However, its effectiveness is highly dependent on atmospheric conditions, and there are ongoing debates about its overall efficacy and potential environmental impacts.

12

Monsoons and Their Mechanism

The Enigmatic Monsoons - Welcome to a fascinating exploration of one of Earth's most powerful and life-giving atmospheric phenomena: the Monsoons. More than just seasonal winds and rains, monsoons are intricate dance performances between land, sea, and atmosphere, shaping the climate, ecology, and very fabric of life across vast stretches of our planet, particularly the Indian subcontinent.

The word "monsoon" itself, derived from the Arabic word "mausim" meaning season, hints at the fundamental characteristic of these weather systems: a seasonal reversal of wind direction. This seemingly simple shift, however, unleashes dramatic transformations. Imagine landscapes parched and brown under the relentless summer sun suddenly bursting into vibrant green as life-sustaining rains arrive. Picture bustling cities and serene villages alike adapting their rhythms to the ebb and flow of the monsoon's arrival and departure. This is the profound influence of the monsoons.

This chapter will delve deep into the heart of these dynamic systems. We will unravel the complex interplay of factors that lead to the birth, progression, and eventual retreat of monsoonal winds and the associated precipitation. We will journey through the scientific understanding of the mechanisms driving these large-scale atmospheric circulations, exploring the roles of differential heating of land and sea, the influence of global pressure belts, the impact of the majestic Himalayas, and the intricate dynamics of atmospheric currents like the Jet Stream.

Beyond the scientific mechanics, we will also explore the profound impact of monsoons on the regions they dominate. For billions of people, particularly in South and Southeast Asia, the monsoon is not just weather; it is life itself. It dictates agricultural cycles, influences cultural traditions, and shapes economic activities. A good monsoon can bring prosperity, while a weak or erratic one can lead to hardship and even disaster.

Furthermore, in our increasingly interconnected world, understanding monsoons is more critical than ever. Climate change is projected to have significant impacts on monsoon patterns, potentially leading to more extreme rainfall events, prolonged dry spells, and shifts in the timing of the seasons. Examining these potential changes and their consequences will also be a crucial aspect of our study.

In the following pages, we will embark on a detailed exploration, covering

- The fundamental definition and characteristics of monsoons.
- The classical and modern theories explaining the monsoon mechanism.
- The different types of monsoons experienced across the globe, with a particular focus on the Indian Monsoon.
- The factors influencing the onset, intensity, and variability of monsoons.
- The socio-economic and environmental impacts of monsoons.
- The challenges and advancements in monsoon prediction.
- The potential influence of climate change on monsoon systems.

Prepare to embark on a captivating journey into the world of monsoons – a powerful testament to the intricate and interconnected nature of our planet's climate system. Let us begin to unravel the mysteries and appreciate the significance of these seasonal winds that hold the key to life for a significant portion of humanity.

12.1. Definition of Monsoons

At its most fundamental, a **monsoon** is a **seasonal reversal of wind direction**, often accompanied by significant changes in precipitation. The term, derived from the Arabic word "mausim" meaning season, aptly describes this cyclical nature.

Traditionally, monsoons were understood as large-scale sea and land breezes caused by the differential heating and cooling of continents and oceans. While this thermal contrast remains a key driving factor, the modern understanding of monsoons recognizes them as planetary-scale phenomena involving the annual migration of the **Intertropical Convergence Zone (ITCZ)** between its northern and southern limits.

In simpler terms, during one part of the year, the prevailing winds may blow consistently from one direction, bringing with them certain weather conditions (like heavy rain from the sea to the land in summer). Then, as the seasons change, these winds shift, often dramatically (blowing from the land to the sea in winter, typically bringing drier conditions).

While often associated with heavy summer rains, it's important to note that a monsoon system technically has both a wet phase and a dry phase corresponding to the seasonal wind shifts.

12.2. Characteristics of Monsoons

Monsoons exhibit several key characteristics that distinguish them from other weather patterns:

- **Seasonal Reversal of Wind Direction:** This is the defining characteristic. The wind patterns shift significantly between seasons, typically by at least 120 degrees. For example, in the Indian Monsoon, winds blow from the southwest during the summer and from the northeast during the winter.
- **Seasonal Precipitation Changes:** The reversal of wind direction is usually accompanied by marked changes in precipitation. Summer monsoons are often associated with heavy and prolonged rainfall as moisture-laden winds blow from the sea onto the land. Winter monsoons, blowing from the land to the sea, are typically dry.
- **Large-Scale Phenomenon:** Monsoons are not localized events like thunderstorms. They are large-scale atmospheric circulations that affect vast geographical areas, sometimes spanning entire continents.
- **Influence of Land-Sea Thermal Contrast:** The differential heating and cooling rates of land and water bodies play a crucial role in driving the monsoon. Land heats up and cools down more quickly than the ocean, creating pressure differences that influence wind flow.
- **Role of the Intertropical Convergence Zone (ITCZ):** The annual migration of the ITCZ, a low-pressure zone near the equator where trade winds converge, is now recognized as a fundamental aspect of monsoon dynamics. The shift in the ITCZ's position influences the direction of winds and the distribution of rainfall.
- **Impact of Topography:** Mountain ranges, like the Himalayas and the Western Ghats in India, can significantly influence monsoon winds and rainfall patterns through orographic lifting (forcing air to rise and condense as it moves over mountains).
- **Wet and Dry Spells:** Within the overall wet season of a monsoon, there can be periods of intense rainfall followed by breaks with little or no rain.This "break" phenomenon is a notable characteristic of some monsoon systems.
- **Variability:** Monsoons are not uniform in their intensity, timing of onset, duration, and spatial distribution of rainfall from year to year. This variability has significant implications, particularly for agriculture.

In essence, monsoons are complex weather systems characterized by a seasonal shift in winds and associated moisture, driven by a combination of thermal contrasts, the movement of global pressure belts, and the influence of geography. They are vital for the water resources and livelihoods of billions of people in the affected regions.

12.3. Classical and Modern Theories on Monsoon Mechanism

The **classical theory**, primarily explains the monsoon as a large-scale version of land and sea breezes, driven by the **differential heating and cooling of land and sea** (Fig. 12.1).

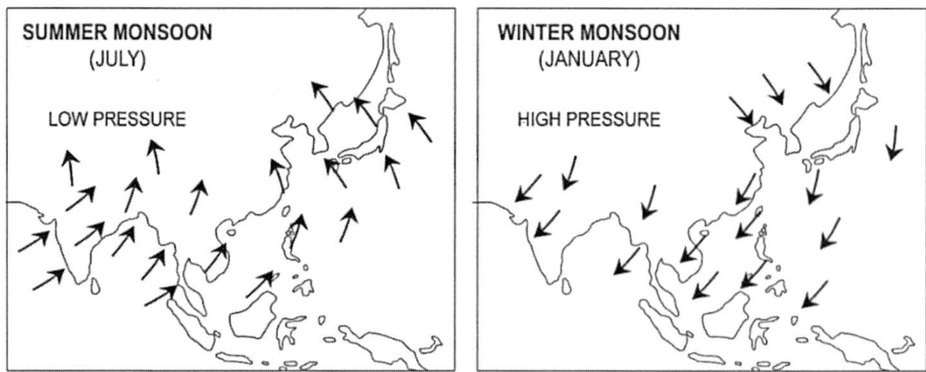

Fig. 12.1: Classical theory on monsoon

12.3.1. Mechanism

- **Summer Monsoon:** During summer, the vast landmass of the Indian subcontinent heats up much more rapidly and intensely than the adjacent Indian Ocean under the influence of direct solar radiation. This intense heating creates a **low-pressure area** over the land. Simultaneously, the ocean remains relatively cooler, resulting in a **high-pressure area** over the sea. Air naturally flows from areas of high pressure to areas of low pressure (Fig. 12.2). Consequently, moisture-laden winds from the high-pressure area over the Indian Ocean are drawn towards the low-pressure area over the subcontinent. As these winds travel over the warm ocean, they pick up significant amounts of moisture. Upon reaching the land and being forced to rise by topography (like the Western Ghats and Himalayas) and thermal convection, the air cools, condenses, and leads to widespread heavy rainfall – the **Southwest Monsoon**.
- **Winter Monsoon:** In winter, the situation reverses. The land cools down much faster than the ocean. This leads to the development of a **high-pressure area** over the cooler landmass and a **low-pressure area**over the relatively warmer ocean. As a result, winds now blow from the high pressure over the land towards the low pressure over the sea. These

winds, originating over the land, are generally dry and hence bring little rainfall to the Indian subcontinent. This is known as the **Northeast Monsoon**. However, as these winds pass over the Bay of Bengal, they can pick up some moisture and cause rainfall over southeastern India, particularly Tamil Nadu.

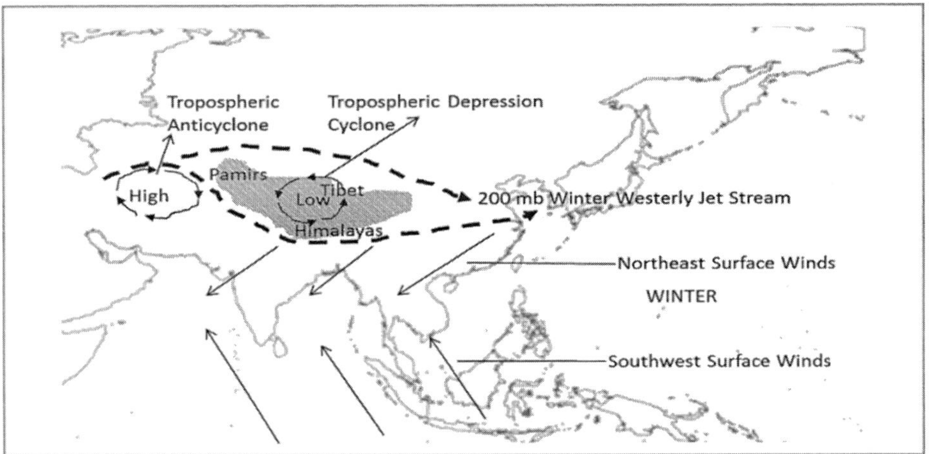

Fig. 12.2: Mechanism of Indian monsoon

12.3.2. Limitations of the Classical Theory

While the thermal contrast between land and sea is undoubtedly a crucial factor, the classical theory fails to explain several aspects of the monsoon, such as:

- The sudden "burst" of the monsoon.
- The intensity and variability of rainfall.
- The influence of upper air circulation.
- The role of geographical features beyond simple land-sea contrast.

12.4. Modern Theories

Modern theories incorporate a more holistic view of the monsoon, recognizing the interplay of various atmospheric and oceanic factors. Some key components of modern monsoon theories include:

- **The Role of the Intertropical Convergence Zone (ITCZ)**
  - The **ITCZ** is a low-pressure zone near the equator where the trade winds from the Northern and Southern Hemispheres converge. It is a zone of rising air, cloud formation, and heavy precipitation (Fig. 12.3).

- The apparent northward and southward movement of the sun throughout the year causes the ITCZ to also shift its position. During summer in the Northern Hemisphere, the ITCZ shifts northwards, reaching over the Indian subcontinent (sometimes as the "monsoon trough" over the Gangetic plains).
- This northward shift of the ITCZ attracts the southeast trade winds of the Southern Hemisphere. After crossing the equator, these winds are deflected to the right due to the Coriolis force, becoming the southwest monsoon winds that bring rain to India.
- In winter, as the sun moves south, the ITCZ also shifts southwards, and the wind patterns reverse.

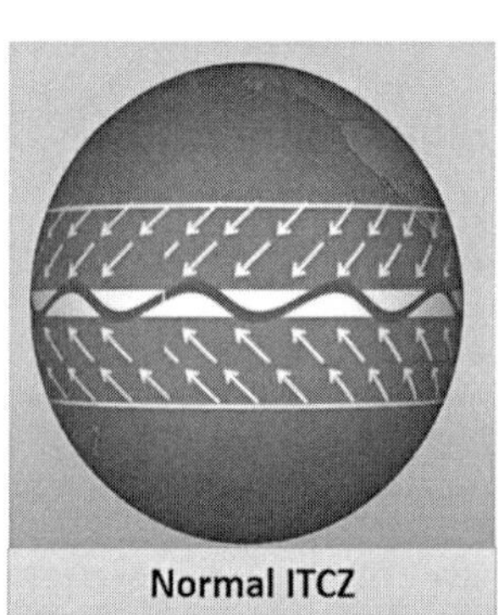

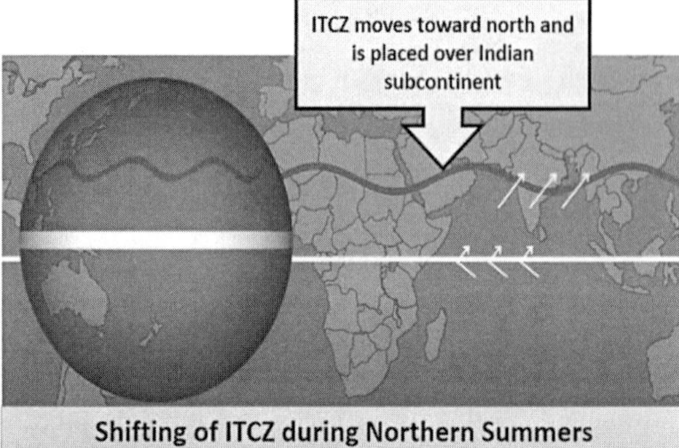

Fig. 12.3: Role of the intertropical convergence zone

- **Influence of the Tibetan Plateau**
  - The **Tibetan Plateau**, a vast elevated landmass to the north of India, plays a significant role in the monsoon mechanism.
  - During summer, the plateau heats up intensely due to its high altitude and receives about 2-3°C more insolation than the surrounding areas. This intense heating creates a strong **low-pressure area** aloft over the plateau.
  - This upper-level low pressure enhances the pressure gradient with the high pressure over the southern Indian Ocean, strengthening the southwest monsoon winds.
    - The heating of the Tibetan Plateau also leads to the development of an upper-level **Tropical Easterly Jet (TEJ)**, which is believed to influence the monsoon circulation and rainfall distribution.

- **The Role of Jet Streams: Jet streams** are fast-flowing, narrow air currents in the upper troposphere. Two jet streams are particularly relevant to the Indian Monsoon:
- **Subtropical Westerly Jet (STWJ):** During winter, the STWJ flows south of the Himalayas, influencing weather patterns over the subcontinent. As summer approaches, the STWJ shifts northwards, generally moving to the north of the Himalayas by early June. This northward shift is considered a crucial precursor to the onset of the southwest monsoon over India.
- **Tropical Easterly Jet (TEJ):** This jet stream develops in the upper troposphere (around 15°N latitude) during the summer months, particularly over peninsular India and Africa. The TEJ is associated with the intense heating of the Tibetan Plateau and is thought to play a role in the intensity of the monsoon.

- **Oceanic Influences**
  - Sea surface temperatures (SSTs) in the Indian Ocean, particularly the Arabian Sea and the Bay of Bengal, significantly impact the moisture content of the monsoon winds.
  - Phenomena like the **Indian Ocean Dipole (IOD)**, characterized by temperature differences between the western and eastern parts of the Indian Ocean, can also influence the monsoon rainfall.
  - Global phenomena like **El Niño-Southern Oscillation (ENSO)** in the Pacific Ocean have also been shown to have teleconnections with the Indian Monsoon, often leading to weaker monsoons during El Niño years.

In conclusion, the modern understanding of the monsoon is that it is a complex phenomenon driven by the seasonal migration of the ITCZ, the thermal influence of the land and the Tibetan Plateau, the dynamics of upper air circulation (jet streams), and oceanic conditions. It's an intricate interplay of these factors that determines the onset, intensity, distribution, and variability of monsoons across the globe.

12.5. Indian Monsoons

Let's refine and expand upon the explanation of the Indian Monsoons, focusing on even greater detail and clarity.

12.5.1. The Enigmatic Indian Monsoons

An In-Depth Exploration: The Indian Monsoon system is far more than just seasonal rain; it's a dramatic, life-sustaining atmospheric upheaval that dictates the rhythm of life across the Indian subcontinent. Its origin is a complex interplay of thermal dynamics, planetary wind patterns, and geographical influences, leading to a distinct cycle of wet and dry seasons.

12.5.2. The Genesis: How the Indian Monsoons Originate

The birth of the Indian Monsoons is a fascinating process driven by a confluence of factors:

- **The Dominant Role of Differential Heating**
    - As summer approaches, the vast landmass of the Indian subcontinent experiences intense solar radiation. Land has a lower specific heat capacity than water, meaning it heats up much faster and to a higher temperature. This intense heating creates a substantial **thermal low-pressure area** at the surface, centered over Northwest India and Pakistan.
    - Simultaneously, the adjacent Indian Ocean, with its higher specific heat capacity, warms up much more slowly. This results in a relatively **high-pressure area** over the ocean.
    - This significant pressure gradient between the land and the sea sets the stage for the large-scale movement of air.
- **The Northward Pilgrimage of the Intertropical Convergence Zone (ITCZ)**
    - The **Intertropical Convergence Zone (ITCZ)** is a low-pressure belt encircling the Earth near the equator, where the northeast and southeast trade winds converge. It's a zone of intense convective activity, cloud formation, and heavy rainfall.
    - Following the apparent northward movement of the sun, the ITCZ also shifts northwards. By summer, it migrates over the Indian subcontinent, often settling around the Gangetic plains. This northward shift intensifies the low-pressure conditions over land, further drawing in oceanic winds.
    - Crucially, the shift of the ITCZ over the subcontinent brings it into contact with the southeast trade winds of the Southern Hemisphere. As these winds cross the equator, the **Coriolis force** (due to Earth's rotation) deflects them to the right in the Northern Hemisphere, causing them to approach India from the southwest.
- **The Profound Influence of the Tibetan Plateau**
    - The elevated **Tibetan Plateau** plays a critical, dynamic role. During summer, this vast plateau heats up intensely due to its altitude and unobstructed solar radiation. The air above the plateau becomes

significantly warmer than the air at the same altitude over the surrounding regions.

- This intense heating creates a strong **thermal low-pressure area in the upper troposphere** above the plateau. This upper-level low pressure enhances the overall low-pressure system over the subcontinent, strengthening the surface pressure gradient and intensifying the inflow of moisture-laden winds from the ocean.
- The heating of the Tibetan Plateau is also linked to the development of the **Tropical Easterly Jet (TEJ)**, a strong upper-level wind current flowing from east to west around 15°N latitude. The TEJ is believed to influence the distribution and intensity of monsoon rainfall.

- **The Shifting Dance of Jet Streams**
  - **Jet streams**, high-altitude, fast-flowing air currents, also play a role in setting the stage for the monsoon.
  - During winter, the **Subtropical Westerly Jet (STWJ)** flows south of the Himalayas, influencing weather patterns over India.
  - As summer approaches, the STWJ retreats northwards, eventually positioning itself to the north of the Himalayas. This northward shift of the STWJ is often seen as a crucial precursor to the onset of the southwest monsoon over India. The absence of the STWJ over the subcontinent allows for the development of the thermal low and the unimpeded northward movement of the ITCZ.
- **The Grand Arrival: The "Burst" of the Monsoon:** The culmination of these processes leads to the dramatic onset of the Southwest Monsoon. Warm, moist air from the Arabian Sea and the Bay of Bengal is drawn towards the low-pressure area over the subcontinent. When these winds encounter the Western Ghats, they are forced to rise, leading to significant orographic rainfall along the west coast. The arrival is often marked by a sudden and substantial increase in rainfall, accompanied by thunderstorms – the dramatic "burst" of the monsoon.
- **The Reign: How the Indian Monsoons Prevail:** Once established, the Indian Monsoon manifests primarily in two distinct phases:
- **The Vigorous Southwest Monsoon (June to September):**
  - This is the period of widespread rainfall across most of India. The southwest monsoon winds, now fully established, are laden with moisture picked up from the warm Indian Ocean (Fig. 12.4).

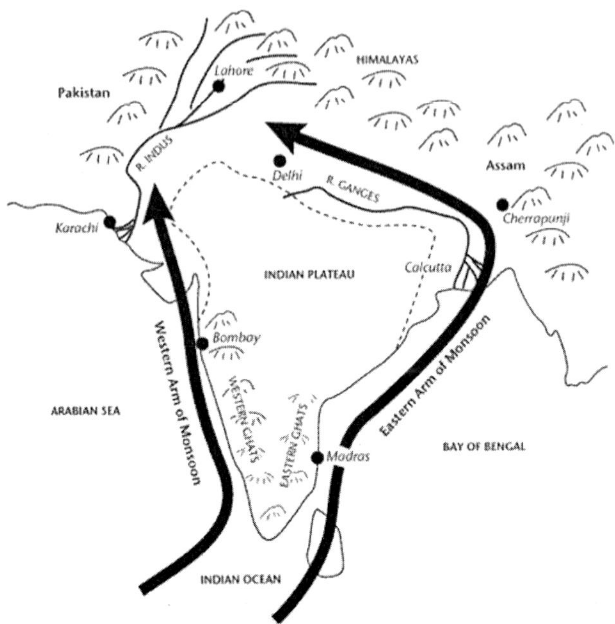

Fig. 12.4: Branches of SW monsoon

The Arabian Sea Branch: This branch strikes the Western Ghats perpendicularly, causing very heavy rainfall on the windward (western) side. As it moves inland, it gets divided into several streams, eventually reaching Northwest India and merging with the Bay of Bengal branch.

- **The Bay of Bengal Branch:** This branch moves northeastwards, bringing copious rainfall to Northeast India, including the Meghalaya plateau (home to Cherrapunji and Mawsynram). A significant portion of this branch then turns westwards along the Gangetic plains, contributing rainfall to central and northern India.
- During this period, the **monsoon trough**, an elongated low-pressure area associated with the ITCZ, oscillates across the Indo-Gangetic plains, influencing the distribution and intensity of rainfall. The position and intensity of this trough are key factors in determining wet and dry spells.
- **The Retreating Northeast Monsoon (October to November):**
  - As autumn sets in, the land begins to cool, and the thermal low-pressure area weakens. A high-pressure area starts to develop over Northwest India.
  - The wind direction reverses, and dry winds begin to blow outwards from the land towards the sea. This is the **Northeast Monsoon**.

- Being land-based, these winds carry little moisture over most of the country. However, as they pass over the Bay of Bengal, they pick up moisture and cause significant rainfall over the southeastern parts of the Indian peninsula, particularly Tamil Nadu, coastal Andhra Pradesh, parts of Kerala, and southeastern Karnataka. This period is the main rainy season for these regions.

12.6. The Hallmarks: Key Characteristics of the Indian Monsoons

The Indian Monsoons are characterized by several distinctive features:

- **Dominant Bimodal Wind System:** The most fundamental characteristic is the near 180-degree seasonal reversal of wind direction.
- **Concentrated Summer Rainfall:** The Southwest Monsoon delivers the vast majority of India's annual precipitation within a relatively short period.
- **Spatial Heterogeneity of Rainfall:** The distribution of monsoon rainfall is highly uneven, influenced by topography, distance from the sea, and the movement of weather systems.
- **Inter-annual Variability:** The timing of onset, total amount of rainfall, its distribution, and the occurrence of breaks vary significantly from year to year, impacting agriculture and water resources.
- **The Phenomenon of "Breaks":** Within the Southwest Monsoon season, there are often periods of reduced or no rainfall lasting for several days or weeks. These "breaks" can be caused by shifts in the monsoon trough or other atmospheric disturbances.
- **The Dramatic "Burst":** The sudden and often intense onset of the Southwest Monsoon is a notable characteristic.
- **Orographic Influence:** Mountain ranges like the Western Ghats and the Himalayas play a crucial role in intercepting monsoon winds and causing significant rainfall on their windward slopes.

12.7. The Lifeline: Importance of the Indian Monsoons

The Indian Monsoons are of paramount importance to the subcontinent:

- **The Agricultural Artery:** Indian agriculture is heavily dependent on monsoon rains for irrigation. The timing and amount of rainfall directly determine the success of the Kharif (summer-sown) crops, which constitute a significant portion of India's agricultural output.
- **Replenishing Water Resources:** The monsoon rains replenish rivers, groundwater aquifers, lakes, and reservoirs, which are vital for

drinking water, irrigation, industrial processes, and hydroelectric power generation.

- **Economic Backbone:** Given the strong link between agriculture and the overall economy, a good monsoon year generally translates to higher economic growth, particularly in rural areas. Conversely, a poor monsoon can have widespread negative economic consequences.
- **Ecological Sustenance:** The monsoon rains are crucial for maintaining the region's diverse ecosystems, supporting forests, wetlands, and biodiversity.
- **Climatic Regulator:** The monsoon circulation plays a significant role in the regional heat and moisture balance.

However, the monsoon's power can also be destructive

- **Devastating Floods:** Intense and prolonged rainfall can lead to widespread flooding, causing loss of life, displacement, and damage to infrastructure and property.
- **Crippling Droughts:** Deficient or delayed monsoon rainfall can result in severe droughts, impacting agriculture, water availability, and livelihoods.

In essence, the Indian Monsoons are a complex, dynamic and absolutely vital weather system. They are the lifeblood of the subcontinent, shaping its climate, economy, and the daily lives of its vast population. Understanding the intricacies of the monsoon is crucial for effective water management, agricultural planning, and disaster preparedness.

12.8. Introduction to the Southwest Monsoon in India

The Southwest Monsoon is the primary rainy season for the Indian subcontinent, typically spanning from June to September. It is a period of dramatic transformation, bringing life-sustaining rains after the hot and dry summer months. The arrival of the southwest monsoon is eagerly awaited across the country, as it replenishes water resources, fuels agriculture, and significantly influences the socio-economic landscape.

- This seasonal shift in weather patterns is characterized by a remarkable change in wind direction. During this period, moisture-laden winds from the Arabian Sea and the Bay of Bengal are drawn towards the low-pressure area that develops over the heated landmass of India. As these winds ascend over various terrains, including the Western Ghats and the Himalayas, they release copious amounts of rainfall.
- The southwest monsoon is not a uniform phenomenon; it exhibits significant spatial and temporal variability. Some regions receive

torrential downpours, while others experience moderate rainfall. The onset, progression, intensity, and duration of the monsoon can vary from year to year, leading to either bountiful harvests or drought-like conditions.

- Understanding the dynamics of the southwest monsoon is crucial for India. It dictates the agricultural calendar, influences water management strategies, and plays a vital role in the overall economy. This introduction will delve into the key aspects of the southwest monsoon, including its origin, the mechanisms driving it, its characteristic features, and its profound importance to the Indian subcontinent. We will explore how this seasonal phenomenon shapes the lives and landscapes of India during these crucial months.

12.9. Date of Onset of Southwest Monsoon

The onset of the Southwest Monsoon over India is a highly anticipated event, marking the transition from the hot, dry pre-monsoon season to a period of widespread rainfall. The India Meteorological Department (IMD) closely monitors various atmospheric and oceanic conditions to predict the date of its arrival.

12.9.1. Date of Onset

- **Normal Onset Date:** The normal date for the onset of the southwest monsoon over Kerala, which marks its entry into the Indian mainland, is **June 1st**, with a standard deviation of about seven days (Fig. 12.5).

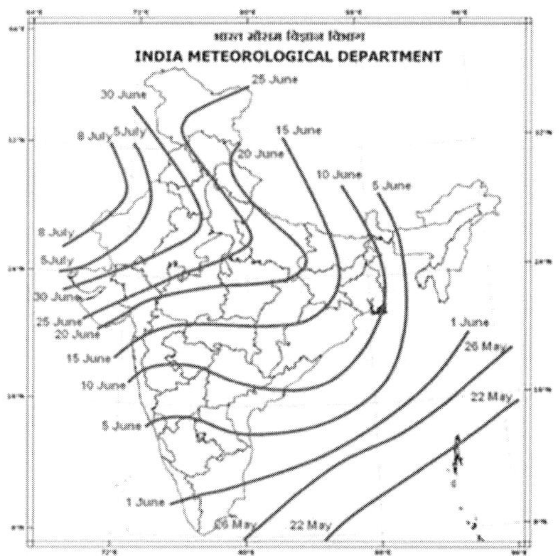

Fig. 12.5: New dates of normal monsoon onset/progress

- **First Point of Entry:** The monsoon typically first arrives over the **Andaman and Nicobar Islands** in the Bay of Bengal, usually around the third week of May (around May 15th to 20th). There's often a lag of about 10 days between its arrival in the Andaman Sea and the Kerala coast.
- **IMD's Prediction for 2025:** For the current year, **2025**, the IMD has predicted an **early onset** of monsoon rain over Kerala, likely around **May 27th**, which is about five days earlier than the usual date.
- **Criteria for Onset Declaration:** The IMD declares the onset of the monsoon over Kerala based on a set of criteria that include:
    - **Rainfall:** At least 60% of 14 designated weather stations in Kerala and Lakshadweep reporting rainfall of 2.5 mm or more for two consecutive days after May 10th.
    - **Wind field:** Westerly winds up to a certain depth in the lower atmosphere and specific wind speeds.
    - **Outgoing Longwave Radiation (OLR):** A measure of energy emitted to space, which should be below a certain threshold, indicating increased cloudiness.

12.9.2. How the Southwest Monsoon Spreads Across India

The southwest monsoon doesn't cover the entire country on a single day. It advances northwards and spreads across India in a phased manner, typically taking about 1 to 1.5 months to cover the entire landmass. The spread occurs in two main branches:

- **The Arabian Sea Branch**
    - This branch first strikes the **Western Ghats** along the southwestern coast of India (Kerala, Karnataka, Goa, Maharashtra). The steep slopes of the Western Ghats cause significant orographic rainfall on the windward side (western side).
    - It then progresses northwards along the west coast, reaching Mumbai generally by around June 10th.
    - Further inland, one stream of this branch moves towards the north-east, reaching Gujarat and eventually merging with the Bay of Bengal branch. Another stream moves along the foothills of the Himalayas.
- **The Bay of Bengal Branch**
    - This branch moves rapidly northeastwards across the Bay of Bengal, striking the coasts of **Myanmar and Bangladesh**, and then entering West Bengal and the northeastern states of India.

- It brings heavy rainfall to Northeast India, including Meghalaya (home to the world's wettest places).
- This branch then turns westwards along the Gangetic plains, gradually spreading across Bihar, Uttar Pradesh, and eventually merging with the Arabian Sea branch around Delhi by late June or early July.

- **Timeline of Spread (Normal Conditions)**
  - **End of May/Early June:** Onset over Kerala and parts of coastal Karnataka.
  - **Mid-June:** Covers most of the west coast, parts of Maharashtra (including Mumbai), and advances into Northeast India. Kolkata usually receives it around June 7th.
  - **Late June/Early July:** Spreads into central India, Gujarat, and reaches Delhi. The two branches (Arabian Sea and Bay of Bengal) merge.
  - **Mid-July:** The monsoon typically covers the entire country by around July 15th.

The advance of the monsoon is often in "surges," with periods of rapid progress followed by slower phases or even temporary halts.10 The topography of the land, the presence of low-pressure systems, and the upper air circulation patterns all influence the speed and extent of its spread.

This year (2025), with the early onset predicted over Kerala, the subsequent spread across the country might also be slightly ahead of the normal schedule, provided there are no significant disruptions in the atmospheric conditions. The IMD will continue to issue updates on the progress of the monsoon as it advances across India.

12.10. Date of Withdrawal of Southwest Monsoon in India

The withdrawal of the Southwest Monsoon over India is a more gradual and prolonged process compared to its relatively rapid onset.1 It begins as the land starts to cool down after the peak summer heat, leading to changes in pressure and wind patterns.

12.10.1. Date of Withdrawal

- **Normal Start Date:** The withdrawal of the southwest monsoon typically begins from the extreme northwestern parts of India (e.g., Rajasthan) around the **first week of September** (Fig. 12.6).

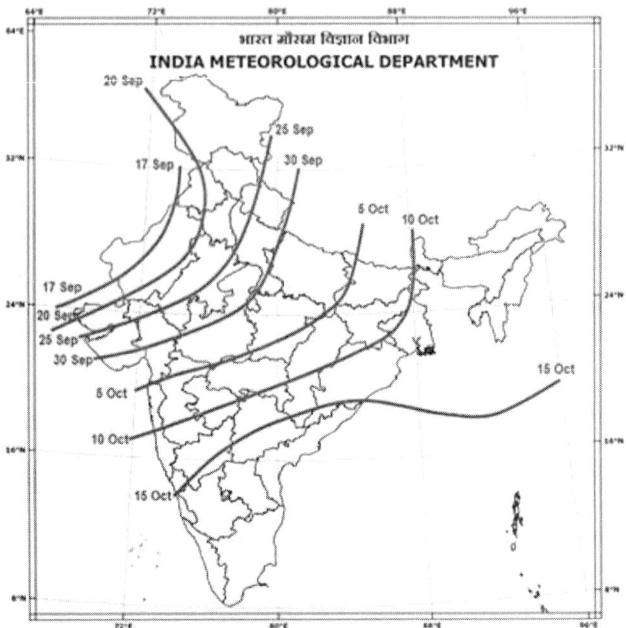

Fig. 12.6: New dates of normal withdrawal monsoon

The India Meteorological Department (IMD) usually sets **September 17th** as the normal date for the commencement of withdrawal from this region.

- **End Date:** The monsoon gradually recedes southwards, and it withdraws completely from the Indian landmass by around **mid-October**. The IMD considers **October 15th** as the normal final withdrawal date from the entire country.
- **Factors Influencing Withdrawal:** The withdrawal is influenced by
  - **Southward movement of the sun:** This leads to the southward shift of the low-pressure trough (associated with the monsoon) from the Gangetic plains.
  - **Establishment of an anticyclone:** The development of an anticyclonic circulation in the lower troposphere over northwestern India, characterized by dry air and sinking motion, signals the end of the monsoon in that region.
  - **Cessation of rainfall:** A continuous period of about five days with no significant rainfall in the northwestern parts.
  - **Reduction in moisture:** A noticeable decrease in moisture content in the lower atmosphere, evident from satellite imagery.

12.10.2. How the Southwest Monsoon Recedes

The withdrawal of the southwest monsoon proceeds generally from northwest to southeast across India:

- **Northwest India (Early September):** The retreat begins from the northwestern states like Rajasthan, Punjab, Haryana, and parts of Gujarat. The air becomes drier, and rainfall ceases.
- **Northern Plains (Mid-September to Early October):** The withdrawal gradually extends to the remaining parts of North India, including Uttar Pradesh, Delhi, and Madhya Pradesh. The monsoon trough weakens and shifts southwards.
- **Central India (Late September to Mid-October):** The monsoon retreats from central parts of India.
- **Peninsular India (October):** The withdrawal from the peninsula is relatively faster. By mid-October, most parts of the peninsula experience a significant reduction in rainfall.
- **Extreme South (Mid to Late October):** The southwest monsoon finally withdraws from the southernmost parts of the peninsula. Around this time, the **Northeast Monsoon** begins to set in over the southeastern peninsula, bringing rainfall to Tamil Nadu, parts of Andhra Pradesh, and Kerala.

12.10.3. Timeline of Withdrawal (Normal Conditions)

- **Early September:** Withdrawal begins from West Rajasthan.
- **Mid-September:** Retreats from most of Northwest India.
- **End of September:** Withdraws from the northern plains and parts of Central India.
- **Mid-October:** Recedes from most of the Indian peninsula.
- **Late October:** Complete withdrawal from the entire country, with the onset of the Northeast Monsoon over the southeast peninsula.

The withdrawal phase is often characterized by clearer skies and a rise in daytime temperatures, while nights become cooler, particularly in North India. This period is sometimes referred to as "October heat" due to the combination of residual moisture and higher temperatures before the full establishment of the winter season.

12.11. Distribution of Southwest Monsoon Rainfall in India

The distribution of Southwest Monsoon rainfall across India is far from uniform, showcasing significant spatial variations due to a complex interplay of geographical factors, wind patterns, and the influence of topography. Here's a detailed look:

12.11.1. Areas of High Rainfall (> 200 cm annually during the SW Monsoon)

- **The Western Ghats (Windward Side):** The Arabian Sea branch of the monsoon strikes the Western Ghats perpendicularly. The orographic effect, where the moist air is forced to rise along the steep slopes, leads to heavy condensation and torrential rainfall on the western side. Some areas here receive over 250 cm of rainfall during this season.
- **Northeast India:** The Bay of Bengal branch of the monsoon flows into Northeast India, getting funneled by the hills. States like Meghalaya (including Mawsynram and Cherrapunji, some of the wettest places on Earth), Assam, Arunachal Pradesh, Nagaland, Manipur, Mizoram, and Tripura receive very high rainfall. The Khasi Hills in Meghalaya significantly enhance rainfall due to their funnel-like shape, trapping the moisture-laden winds.
- **Sub-Himalayan regions:** Areas along the foothills of the Himalayas also receive substantial rainfall as the Bay of Bengal branch moves westward and is forced to rise.

12.11.2. Areas of Medium Rainfall (100 - 200 cm annually during the SW Monsoon)

- **Southern parts of Gujarat:** Receives a moderate amount of rainfall from the Arabian Sea branch.
- **East Tamil Nadu:** While the southwest monsoon is not the primary rainy season for Tamil Nadu, it does receive some rainfall during this period.
- **Northeastern Peninsula:** This includes Odisha, Jharkhand, Bihar, and eastern Madhya Pradesh, which receive rainfall from the Bay of Bengal branch as it moves inland.
- **Northern Ganga Plain (along the sub-Himalayas):** As the Bay of Bengal branch moves westward.
- **Coastal Karnataka:** Receives good rainfall as part of the Arabian Sea branch's influence.

12.11.3. Areas of Low Rainfall (50 - 100 cm annually during the SW Monsoon)

- **Western Uttar Pradesh, Delhi, Haryana, Punjab:** These areas receive rainfall as the monsoon progresses westward, but the moisture content tends to decrease with distance from the Bay of Bengal. The merging of the Arabian Sea and Bay of Bengal branches also contributes.

- **Eastern Rajasthan:** Receives relatively less rainfall from the Arabian Sea branch as it moves parallel to the Aravalli Range. The Aravallis, running southwest-northeast, do not effectively block the monsoon winds to cause significant orographic lifting and rainfall in this region.
- **Gujarat (excluding the southern parts):** The northern parts and Kutch experience lower rainfall from the Arabian Sea branch, which moves parallel to the Aravallis.
- **Deccan Plateau:** Much of the interior Deccan Plateau receives less rainfall as it lies in the rain shadow region of the Western Ghats for the Arabian Sea branch. The monsoon winds lose much of their moisture by the time they cross the Ghats.

12.11.4. Areas of Inadequate Rainfall (< 50 cm annually during the SW Monsoon)

- **Parts of the Peninsula (especially leeward side of Western Ghats):** Regions in Maharashtra, Karnataka, and Andhra Pradesh that lie on the eastern side of the Western Ghats fall in the rain shadow zone and receive very scanty rainfall from the Arabian Sea branch. For example, Mahabaleshwar on the windward side receives over 600 cm, while Pune on the leeward side gets only around 70 cm.
- **Western Rajasthan:** This arid region receives very little rainfall from the Arabian Sea branch, which passes by without significant obstruction.
- **Ladakh:** Being a high-altitude cold desert, it receives minimal rainfall due to its location in the rain shadow of the Himalayas.

12.11.5. Factors Influencing the Distribution

- **Orographic Lift:** Mountain ranges like the Western Ghats and the Himalayas force moisture-laden winds to rise, causing heavy rainfall on the windward side. The leeward side experiences a rain shadow effect with much less rainfall.
- **Distance from the Sea:** Generally, rainfall decreases with increasing distance from the coast due to the gradual depletion of moisture in the winds.
- **Monsoon Depressions:** Low-pressure systems that form in the Bay of Bengal and the Arabian Sea move inland, significantly influencing the spatial distribution of rainfall along their track.

- **Wind Direction and Flow:** The direction of the monsoon winds and the patterns of air circulation determine which areas receive more direct and moisture-rich winds.
- **Topography:** Besides mountains, other topographical features can influence local rainfall patterns.
- **Location of the Monsoon Trough (ITCZ):** The position and movement of the monsoon trough across the Indo-Gangetic plains significantly affect rainfall distribution.

In summary, the distribution of southwest monsoon rainfall in India is highly varied, with the western coast and Northeast receiving the highest amounts, followed by the sub-Himalayan regions and parts of the central plains. Rainfall generally decreases as one moves westwards across the northern plains and into the leeward side of the Western Ghats and the arid regions of the northwest.

12.12. Break in the Monsoon: A Detailed Explanation

A **break in the monsoon** refers to a period during the Southwest Monsoon season (typically July and August) when there is a **significant reduction or complete cessation of rainfall over most parts of India**, particularly over the central and northern plains and the west coast. This dry spell occurs within the overall wet monsoon season, interspersed between periods of active rainfall. While the plains might experience a lull, rainfall often increases simultaneously along the Himalayan foothills, Northeast India, and sometimes parts of the southern peninsula.

Think of the monsoon not as continuous, uninterrupted rain, but as a period characterized by alternating wet spells and these drier "breaks."

12.12.1. What Causes Break Monsoon Conditions?

The primary cause of a break in the monsoon is the **northward shift of the monsoon trough** (the low-pressure zone associated with the ITCZ) closer to the Himalayan foothills. Here's a breakdown of the mechanism:

- **Normal Position of the Monsoon Trough:** Typically, during the peak monsoon months, the monsoon trough extends from Rajasthan in the west to West Bengal in the east, across the Indo-Gangetic plains. This trough attracts moisture-laden winds from the Arabian Sea and the Bay of Bengal, leading to widespread rainfall across central and northern India.
- **Northward Shift of the Monsoon Trough:** When the monsoon trough shifts northwards and aligns itself along the foothills of the Himalayas, a significant change in the rainfall pattern occurs:

  - **Cessation of Rainfall over the Plains:** As the low-pressure system moves north, the plains of northern and central India experience a weakening of the monsoon circulation and a drastic reduction or complete stop in rainfall. The moisture-bearing winds are no longer effectively drawn over these areas.
  - **Increased Rainfall in the Himalayas and Northeast:** With the monsoon trough closer to the Himalayas, the monsoon winds are forced to ascend the steep slopes, leading to enhanced orographic rainfall in the Himalayan region and the northeastern states. This can sometimes result in floods in these areas.
- Other factors that can contribute to breaks in the monsoon include:
  - **Absence of Monsoon Depressions:** The formation and movement of low-pressure systems or monsoon depressions over the Bay of Bengal are crucial for maintaining active monsoon conditions. If these depressions are infrequent or do not follow their usual track, it can lead to a break.
  - **Westerly Winds over the West Coast:** When winds along the west coast become parallel to the coastline instead of perpendicular, they carry less moisture inland, leading to dry spells in the region.
  - **Global Teleconnections:** Phenomena like El Niño-Southern Oscillation (ENSO) in the Pacific Ocean can influence the position and strength of the monsoon trough, indirectly contributing to break conditions.

12.12.2. Impact of Break Monsoons

Breaks in the monsoon can have significant and varied impacts:

- **Agriculture:** The most severely affected sector is agriculture, particularly rain-fed crops. Prolonged dry spells during crucial growth stages can lead to
  - **Moisture stress in crops:** Wilting and stunted growth due to lack of water.
  - **Reduced yields:** If the break is long and not followed by sufficient rainfall, it can significantly decrease agricultural productivity.
  - **Delayed sowing for the next season:** If the withdrawal of the monsoon is also delayed after a long break, it can disrupt agricultural schedules.
- **Water Resources:** Breaks can lead to a halt in the replenishment of rivers, reservoirs, and groundwater, potentially leading to water scarcity if the break is prolonged.

- **Economy:** Given the dependence of the Indian economy on agriculture, a significant break in the monsoon that impacts crop yields can have broader economic repercussions, affecting rural incomes and overall growth.
- **Paradoxical Situations:** As mentioned earlier, while the plains suffer from a lack of rain, the Himalayan and northeastern regions might experience floods due to concentrated rainfall, leading to a paradoxical situation of drought in one part of the country and floods in another.
- **Ecosystems:** Prolonged dry spells can also impact natural vegetation and local ecosystems.

In summary, a break in the monsoon is a temporary but significant disruption in the typical rainfall pattern of the Southwest Monsoon, primarily caused by the northward shift of the monsoon trough. It leads to dry spells over central and northern India while often increasing rainfall in the Himalayan and northeastern regions, with potentially adverse impacts on agriculture, water resources, and the economy.

12.13. Northeast Monsoon

The **Northeast Monsoon**, also known as the **winter monsoon** or the **retreating monsoon**, is a distinct weather phenomenon affecting the Indian subcontinent, primarily during the months of **October to December**.While it contributes a smaller portion to India's overall annual rainfall compared to the Southwest Monsoon, it is critically important for the southeastern parts of the peninsula (Fig. 12.7).

12.13.1. Formation of the Northeast Monsoon

The formation of the Northeast Monsoon is linked to the changes in atmospheric circulation that occur as the summer heating over the Indian landmass subsides and winter approaches:

- **Cooling of the Landmass:** After the Southwest Monsoon season, the Indian landmass begins to cool down more rapidly than the surrounding Indian Ocean. This cooling leads to the development of a **high-pressure area** over northern India.
- **Southward Shift of the ITCZ:** As the sun moves southwards towards the Tropic of Capricorn, the **Intertropical Convergence Zone (ITCZ)** also shifts south. The low-pressure trough associated with the monsoon weakens over the northern plains and gradually moves southward.
- **Reversal of Winds:** Due to the high pressure over the land and relatively lower pressure over the Bay of Bengal, the wind direction reverses.

Cooler, drier air starts blowing outwards from the northeast across the land towards the sea. These are the **northeast trade winds**.

- **Moisture Pickup over the Bay of Bengal:** As these northeasterly winds travel over the warm waters of the Bay of Bengal, they pick up significant amounts of moisture. This moisture-laden air then brings rainfall to the eastern coast of South India.

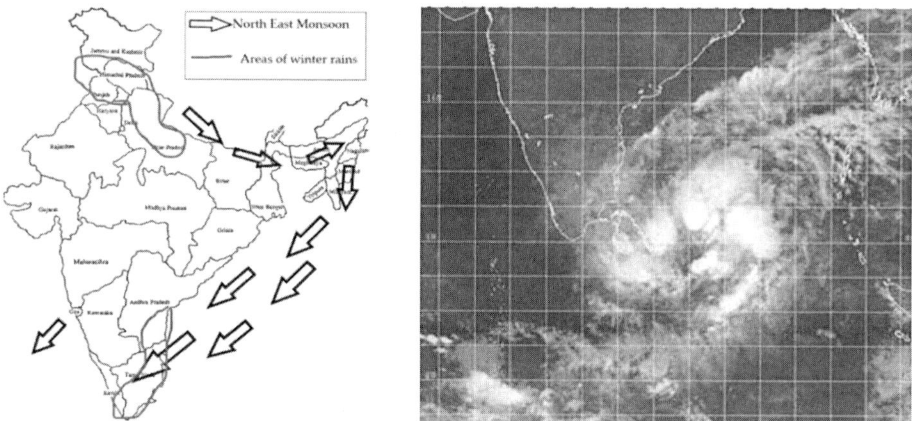

Fig. 12.7: Northeast monsoon

12.13.2. Spread of the Northeast Monsoon

Unlike the relatively well-defined onset and spread of the Southwest Monsoon from south to north, the onset of the Northeast Monsoon is more gradual and less dramatic. Its influence is largely confined to the southeastern peninsula:

- **Commencement:** The establishment of the northeasterly winds over the Bay of Bengal and the onset of rainfall along the Tamil Nadu coast usually marks the beginning of the Northeast Monsoon, typically around **mid-October**.
- **Limited Spread:** The influence of the Northeast Monsoon is primarily felt over **Tamil Nadu, coastal Andhra Pradesh, parts of Kerala, and south interior Karnataka**. It doesn't extend significantly into the interior of the country or the northern plains.
- **Rainfall Mechanism:** The rainfall associated with the Northeast Monsoon is often caused by:
  - **Easterly waves:** These are undulations in the tropical easterly flow that can bring rainfall.
  - **Low-pressure systems, depressions, and tropical cyclones:** The post-monsoon season (October-December) is the peak time for the

formation of tropical cyclones in the Bay of Bengal. These systems can bring heavy and widespread rainfall to the eastern coast as they make landfall. Some of these cyclones can even cross the southern peninsula and emerge into the Arabian Sea.

12.13.3. Areas Under the Influence of the Northeast Monsoon in India

The Northeast Monsoon has a significant impact on the following regions:

- **Tamil Nadu:** This state is the **primary beneficiary** of the Northeast Monsoon, receiving about **40-60% of its annual rainfall** during this season. For many parts of Tamil Nadu, this is the main rainy season, crucial for agriculture (especially paddy, millets, and groundnut cultivation) and replenishing water reservoirs. The coastal districts often receive the heaviest rainfall, sometimes associated with cyclonic activity.
- **Coastal Andhra Pradesh:** This region also receives significant rainfall from the Northeast Monsoon, accounting for about **30-40% of its annual rainfall** in some areas, particularly the southern coastal districts. The rainfall is vital for agriculture and water management in this region.
- **Parts of Kerala:** While Kerala receives the bulk of its rainfall from the Southwest Monsoon, the southern parts of the state experience a **secondary rainy season** during the Northeast Monsoon. This rainfall is important for winter crops and maintaining water levels.
- **South Interior Karnataka:** The southeastern parts of Karnataka also receive some rainfall during the Northeast Monsoon. While not as high as in Tamil Nadu or coastal Andhra Pradesh, this rainfall is still beneficial for agriculture and water resources in this region.

12.13.4. Other Areas with Marginal Influence

- **Rayalaseema (Andhra Pradesh):** Experiences some spillover rainfall from the Northeast Monsoon.
- **Puducherry and Karaikal:** Being geographically close to Tamil Nadu, these Union Territories also receive substantial rainfall during this season.

12.13.5. Areas Largely Unaffected

The Northeast Monsoon has minimal to no direct impact on most other parts of India, including the northern plains, central India, the west coast (except for some parts of Kerala), and Northeast India (which receives its main rainfall during the Southwest Monsoon).

In essence, the Northeast Monsoon is a crucial, albeit geographically limited, rainy season for the southeastern peninsula of India, particularly Tamil Nadu. It forms due to the reversal of winds after the summer monsoon, picking up moisture from the Bay of Bengal and bringing vital precipitation to these regions.

12.14. Distribution of Northeast Monsoon Rainfall in India

The **Northeast Monsoon** primarily influences the southeastern parts of the Indian peninsula during the months of **October, November, and December**. Let's discuss its distribution with this timeframe in mind:

12.14.1. October

- As the Southwest Monsoon withdraws, the shift in wind patterns becomes noticeable. By mid-October, the northeasterly winds start establishing themselves over the Bay of Bengal.
- **Tamil Nadu** and **coastal Andhra Pradesh** begin to experience an increase in rainfall. This is often the initial phase of their main rainy season. The rainfall during this month can be quite variable, sometimes starting with lighter showers and gradually intensifying.
- **Kerala** and **south interior Karnataka** also start receiving some rainfall, marking the beginning of their secondary rainy season.
- The formation of **tropical cyclones** in the Bay of Bengal becomes more frequent in October. These cyclones can bring heavy and widespread rainfall, along with strong winds, to the eastern coast, particularly affecting Tamil Nadu, Andhra Pradesh, and sometimes Odisha.

12.14.2. November

- November is typically the **peak month** for the Northeast Monsoon in terms of rainfall amount and its impact on the southeastern peninsula.
- **Tamil Nadu** receives its highest rainfall during this month. Most parts of the state experience consistent and often heavy rainfall.
- **Coastal Andhra Pradesh** also continues to receive significant rainfall, and the risk of cyclones impacting the coast remains high.
- **Kerala** and **south interior Karnataka** continue to receive moderate rainfall.
- The influence of the Northeast Monsoon generally does not extend far inland or to the northern parts of India during this period. The rest of the country experiences largely dry and pleasant weather as the transition to the winter season progresses.

12.14.3. December

- The intensity of the Northeast Monsoon starts to decrease in December.
- Rainfall in **Tamil Nadu** and **coastal Andhra Pradesh** gradually reduces compared to November, but significant rainfall events can still occur, especially in the first half of the month.
- **Kerala** and **south interior Karnataka** also see a further reduction in rainfall.
- By the end of December, the Northeast Monsoon season effectively comes to an end, and the weather over the southeastern peninsula starts to become drier as the winter conditions become more established.

12.14.4. Summary of Regional Distribution over the Period

- **Tamil Nadu (October-December):** Receives the bulk (around 40-60%) of its annual rainfall, with November being the rainiest month. Coastal districts often get the heaviest.
- **Coastal Andhra Pradesh (October-December):** Experiences significant rainfall (30-40% of annual), with the risk of cyclonic rainfall being higher in October and November.
- **Kerala (October-December):** Receives its secondary rainy season, with moderate rainfall, mainly in the southern parts.
- **South Interior Karnataka (October-December):** Gets some rainfall, though less than the eastern coastal areas.
- **Rest of India (October-December):** Largely unaffected by the Northeast Monsoon, experiencing the transition towards the winter season with generally dry weather.

In essence, the distribution of the Northeast Monsoon's rainfall is heavily concentrated over the southeastern peninsula during October to December, with a peak in November. Tamil Nadu and coastal Andhra Pradesh are the primary beneficiaries, while Kerala and south interior Karnataka receive a lesser but still important amount. The influence diminishes significantly as one moves inland or northwards.

12.15. Cyclones and Depressions during Northeast Monsoon Season

The Bay of Bengal is a breeding ground for tropical disturbances, some of which intensify into depressions and cyclones, particularly during the **post-monsoon season (October-December)**, which aligns with the Northeast Monsoon. Several factors contribute to this: warm sea surface temperatures, lower vertical wind shear compared to the peak monsoon, and the remnants of typhoons from the Pacific sometimes entering the Bay of Bengal.

12.15.1. Depressions

A depression is the first stage in the intensification of a tropical cyclone. It is characterized by:

- **Low-pressure area:** A defined low-pressure center.
- **Organized convection:** Thunderstorms that show some degree of organization around the center.
- **Wind speed:** Maximum sustained surface winds up to **62 km/h (38 mph or 33 knots)**.

  During the Northeast Monsoon season, depressions frequently form over the Bay of Bengal. While they are weaker than cyclones, they can still bring:
- **Heavy rainfall:** Capable of causing flooding in coastal and inland areas.
- **Gusty winds:** Leading to minor damage to weak structures and trees.
- **Rough seas:** Affecting fishing and maritime activities.

Many of the rainfall events associated with the Northeast Monsoon, especially in Tamil Nadu and Andhra Pradesh, are linked to the passage of these depressions. Some depressions intensify further into tropical cyclones.

12.15.2. Cyclones

A tropical cyclone is a more intense weather system that develops from a tropical depression when certain favorable conditions persist. It is characterized by:

- **Well-defined low-pressure center (the eye):** A region of relatively calm weather at the center.
- **Organized and intense convection:** Bands of strong thunderstorms spiraling around the eye (eyewall being the most intense).
- **Wind speed:** Maximum sustained surface winds exceeding **63 km/h (39 mph or 34 knots)**. Depending on the intensity, they can be further classified into tropical storms, severe cyclonic storms, very severe cyclonic storms, and super cyclonic storms.

The post-monsoon season (October-December) is one of the **peak times for cyclone formation in the Bay of Bengal**. These cyclones can have a significant impact on the eastern coast of India and neighboring countries like Bangladesh and Myanmar.

12.15.3. Key aspects of cyclones during the Northeast Monsoon

- **Frequency:** The Bay of Bengal experiences a higher frequency of tropical cyclones compared to the Arabian Sea. On average, a few cyclones form in the Bay of Bengal during this season.

- **Tracks:** Cyclones formed during this period often move west or northwest initially, posing a threat to the eastern coast of India (Tamil Nadu, Andhra Pradesh, Odisha, West Bengal). Some may recurve northeastwards.
- **Intensity:** These cyclones can range from tropical storms to very severe cyclonic storms, capable of causing widespread destruction.
- **Impact:** The impact can be severe, including
  - **Storm surges:** Abnormal rise in sea level inundating coastal areas.
  - **High winds:** Causing damage to infrastructure, homes, and vegetation.
  - **Torrential rainfall:** Leading to widespread flooding.

12.15.4. Examples (Hypothetical for the 2025 Northeast Monsoon)

Given the current date (May 15, 2025), we cannot provide specific examples of cyclones that will occur during the upcoming Northeast Monsoon. However, historically, this period has seen the formation of significant cyclones. For instance, in November 2024, Cyclone Michaung impacted Tamil Nadu and Andhra Pradesh.

In the upcoming Northeast Monsoon (October-December 2025), it is plausible that:

- Several low-pressure systems and depressions will form over the Bay of Bengal, bringing periods of enhanced rainfall to the southeastern peninsula.
- At least one or two of these systems might intensify into tropical cyclones, potentially making landfall along the eastern coast, bringing the associated hazards of strong winds, heavy rain, and storm surges to vulnerable coastal districts.

The India Meteorological Department (IMD) closely monitors the Bay of Bengal for the development of such systems and issues timely warnings.

In summary, the Northeast Monsoon season in India is characterized by the formation of depressions and cyclones in the Bay of Bengal, which are crucial for the rainfall in the southeastern peninsula but also pose a significant risk of severe weather events along the eastern coastline.

12.16. Distribution of Annual Rainfall in India

The average annual rainfall of India is about **118 cm**, but this figure hides a vast disparity in its distribution across different regions. Several factors such as topography, distance from the sea, prevailing wind direction, and the location of mountain ranges significantly influence how much rainfall different parts of India receive.

Here's a detailed look at the distribution of annual rainfall in India

12.16.1. Zones of Very High Rainfall (Above 200 cm annually)

- **The Western Ghats (Windward Side):** The western slopes of the Western Ghats, intercepting the moisture-laden Arabian Sea branch of the Southwest Monsoon, receive exceptionally heavy rainfall, often exceeding **250 cm**.Areas like the Konkan coast, parts of Karnataka and Kerala within the Ghats experience this. The orographic effect is the primary reason here.
- **Northeast India:** This region, comprising states like Meghalaya, Assam, Arunachal Pradesh, Nagaland, Manipur, Mizoram, and Tripura, receives very high rainfall primarily from the Bay of Bengal branch of the Southwest Monsoon.
    - **Meghalaya:** Contains some of the wettest places on Earth, such as **Mawsynram** and **Cherrapunji**, which can receive over **1000 cm** of rainfall annually. The unique funnel shape of the Khasi Hills enhances the concentration of moisture-laden winds.
    - **Sub-Himalayan West Bengal and Sikkim:** These areas also receive high rainfall due to their proximity to the Himalayas, which force the monsoon winds to rise.

12.16.2. Zones of High Rainfall (100 - 200 cm annually)

- **Eastern India:** This includes parts of West Bengal, Odisha, Bihar, and Jharkhand, which receive substantial rainfall from the Bay of Bengal branch of the Southwest Monsoon as it moves inland.
- **Coastal Plains:** Areas along the western and eastern coastal plains (excluding Tamil Nadu coast during the SW monsoon) generally receive rainfall in this range.
- **Parts of the Northern Plains:** Areas along the sub-Himalayan belt of the northern Ganga plain also fall into this category.
- **Southern Gujarat:** Receives a good amount of rainfall from the Arabian Sea branch.

12.16.3. Zones of Moderate Rainfall (50 - 100 cm annually)

- **Central India:** Much of Madhya Pradesh and parts of Chhattisgarh receive moderate rainfall.
- **Deccan Plateau:** The rain shadow region to the east of the Western Ghats, covering parts of Maharashtra, Karnataka, and Andhra Pradesh, receives comparatively less rainfall.

- **Parts of the Northern Plains:** Western Uttar Pradesh, Haryana, and Punjab receive rainfall in this range as the monsoon weakens moving westward.
- **Eastern Rajasthan:** Receives less rainfall as the Arabian Sea branch moves parallel to the Aravalli Range, offering little obstruction.

12.16.4. Zones of Low Rainfall (Less than 50 cm annually)

- **Western Rajasthan:** The arid and semi-arid regions of western Rajasthan receive very scanty rainfall, often less than **25 cm**. The Arabian Sea branch of the monsoon flows parallel to the Aravalli hills, and there are no significant barriers to cause orographic lift.
- **Parts of Gujarat:** The northern parts of Gujarat, including the Kutch region, also receive low rainfall.
- **Leeward side of the Western Ghats:** A significant area of the Deccan Plateau falls in the rain shadow region of the Western Ghats and thus receives very little rainfall.
- **Ladakh:** Being a cold desert situated in the rain shadow of the Himalayas, Ladakh receives extremely low rainfall.

12.16.5. Factors Explaining the Uneven Distribution

- **Topography:** The presence of mountain ranges like the Western Ghats and the Himalayas plays a crucial role in intercepting monsoon winds and causing orographic rainfall on the windward side, while creating rain shadow regions on the leeward side.
- **Distance from the Sea:** Coastal areas generally receive more rainfall than the interior regions due to their proximity to moisture sources.
- **Direction of Monsoon Winds:** Areas that lie directly in the path of the moisture-laden monsoon winds receive more rainfall. For example, the southwest monsoon winds directly hit the Western Ghats and Northeast India.
- **Influence of Ocean Currents:** Sea surface temperatures influenced by ocean currents can affect the amount of moisture in the air.
- **Low-Pressure Systems:** The formation and movement of monsoon depressions and cyclones in the Bay of Bengal and the Arabian Sea significantly influence the distribution of rainfall. Areas along their track receive heavier rainfall.

In conclusion, the distribution of annual rainfall in India is highly uneven, ranging from some of the wettest places on Earth to arid deserts, all within the same country. This diversity is primarily due to the complex interplay of geographical and meteorological factors that influence the behavior of the monsoon winds and the occurrence of precipitation.

12.17. Importance of The Monsoons: India's Life Force - A Nation's Dependence

The monsoons are not just a seasonal weather event for India; they are the very engine that drives its economy, sustains its vast population, and shapes its cultural and environmental landscape. Their influence is so profound that the annual performance of the monsoon is a matter of national concern, impacting everything from agricultural yields to the stock market.

12.17.1.. The Primacy of Agriculture - Feeding a Billion-Plus

- **The Unirrigated Majority:** While irrigation infrastructure has expanded, a significant portion of India's agricultural land remains rain-fed. For these vast tracts, the Southwest Monsoon is the sole and critical source of water for cultivation. States like Maharashtra, parts of Karnataka, and even areas within Telangana rely heavily on monsoon rains for their primary cropping seasons.
- **Key Crop Dependence:** Staple crops like rice, which feeds a massive population across the country (including the rice-growing regions of Telangana), are almost entirely dependent on monsoon rainfall. Similarly, other crucial crops like pulses, oilseeds, cotton, and sugarcane rely heavily on the monsoon for their growth cycle. Any deviation in the timing, amount, or distribution of monsoon rainfall directly threatens food security for the entire nation.
- **Livestock and Allied Sectors:** Beyond crops, the monsoon also impacts livestock by ensuring the availability of fodder and drinking water in rural areas across India. Dairy farming, a significant contributor to the rural economy, also indirectly benefits from a good monsoon.
- **National Economic Indicator:** The agricultural sector's performance, intrinsically linked to the monsoon, has a significant bearing on India's Gross Domestic Product (GDP). Years of good monsoon rainfall often correlate with higher economic growth rates, while drought years can lead to economic slowdowns affecting all states, including a rapidly growing one like Telangana.

12.17.2. The Lifeline of Water Resources - Sustaining a Thirsty Nation

- **River Systems:** The monsoon rains are the primary source of water for India's major river systems, which crisscross the country, providing water for irrigation, drinking, and industrial use from the Himalayas in the north to the peninsular rivers in the south. The Ganga, Brahmaputra, Godavari, Krishna, and Kaveri basins all rely on the monsoon to replenish their flows.

- **Groundwater Recharge:** In many parts of India, including areas of Telangana facing water stress, groundwater is a critical resource. The monsoon is the primary mechanism for recharging these underground aquifers, ensuring long-term water security for both rural and urban populations.
- **Hydroelectric Power:** India's hydroelectric power generation capacity is significantly dependent on the water levels in reservoirs, which are filled by monsoon rains. Consistent monsoon rainfall ensures a stable supply of electricity, crucial for industrial and domestic consumption across the nation.

12.17.3. The Broader Economic Impact - Beyond the Farm Gate

- **Rural Economy and Demand:** The majority of India's population still resides in rural areas, and their economic well-being is closely tied to agriculture. A good monsoon boosts rural incomes, leading to increased purchasing power and driving demand for a wide range of goods and services, benefiting industries and businesses nationwide. For example, strong rural demand impacts the sales of everything from motorcycles in Telangana's villages to consumer goods in metropolitan cities.
- **Inflation Management:** Food inflation is a major concern in India. A favorable monsoon, leading to ample agricultural production, helps stabilize food prices, thus controlling overall inflation and providing relief to consumers across all economic strata throughout the country.
- **Industrial Linkages:** Many industries in India, such as textiles (cotton), sugar, food processing, and breweries, depend directly or indirectly on agricultural output, which in turn is heavily reliant on the monsoon. A good monsoon fuels these industries, contributing to overall economic growth.

12.17.4. Environmental Stewardship - Nourishing India's Ecology

- **Biodiversity and Ecosystems:** The monsoon is fundamental to maintaining India's rich and diverse ecosystems, from the rainforests of the Western Ghats and Northeast to the deciduous forests of the central plains. It provides the necessary moisture for vegetation, sustains wildlife, and supports the delicate balance of these natural habitats across the country.

- **Soil Health and Fertility:** Monsoon rains are essential for maintaining soil moisture, which is crucial for plant growth and overall soil health across India's varied agro-climatic zones.

12.17.5. Socio-Cultural Integration - The Rhythm of Life

- **Festivals and Traditions:** The anticipation and arrival of the monsoon are deeply embedded in India's cultural fabric. Numerous festivals across different states celebrate the rains, signifying their importance in the collective consciousness. From the Teej festival in the north to Onam in Kerala, the monsoon's influence is woven into the social and cultural life of the nation.
- **Art, Literature, and Folklore:** The monsoon has been a constant source of inspiration for Indian artists, writers, and poets for centuries, symbolizing life, renewal, and the power of nature in the country's rich artistic heritage.

12.17.6. The National Vulnerability: The Flip Side

It is crucial to acknowledge that the Indian economy and society are also highly susceptible to the vagaries of the monsoon.

- **Droughts:** A weak or delayed monsoon can lead to widespread droughts, causing immense hardship to farmers, depleting water resources, impacting power generation, and potentially leading to migration and social unrest across affected regions.
- **Floods:** Conversely, excessive and concentrated monsoon rainfall can trigger devastating floods, causing loss of life, damage to infrastructure, displacement of populations, and significant economic losses in various parts of the country, including coastal areas prone to cyclones during the monsoon and post-monsoon periods.

In conclusion, the monsoon is far more than just a weather phenomenon for India. It is the **linchpin of its economy**, the **sustainer of its water resources**, a **vital force for its environment**, and a deeply ingrained aspect of its **culture**. Its performance has direct and far-reaching consequences for the well-being of over a billion people, making its accurate prediction and effective management a matter of paramount national importance. From the fields of down south to the snow-capped Himalayas, the rhythm of life in India beats in sync with the arrival and departure of the monsoons.

13

Weather Hazards

Weather, while often benign and essential for life, can also manifest in powerful and destructive forms, posing significant threats to human life, property, and the environment. This chapter, "Weather Hazards," will delve into the detailed nature of these potentially dangerous atmospheric phenomena.

We will explore a range of significant weather hazards, including severe storms such as thunderstorms, tornadoes, and hailstorms, which can bring intense winds, torrential rainfall, and damaging projectiles. We will also examine the characteristics and impacts of tropical cyclones (hurricanes, typhoons), known for their devastating winds, storm surges, and widespread flooding. Furthermore, we will discuss the dangers associated with **extreme temperature events, including heat waves, characterized by prolonged periods of excessively hot weather, and cold waves, marked by extended periods of dangerously cold temperatures.** We will also cover the challenges posed by heavy snowfall, blizzards, and ice storms.

For each hazard, we will investigate the meteorological conditions that lead to its development, the typical characteristics and intensity scales used to classify it, and the primary impacts it can have on society and the natural world. Understanding the formation, behavior, and potential consequences of these weather hazards is crucial for effective forecasting, issuing timely warnings, and implementing mitigation strategies to enhance safety and resilience.

Join us as we embark on a detailed exploration of the forces of nature that can transform familiar weather into formidable hazards.

13.1. Floods

A **flood** is an overflow of water that submerges land and crops. This can happen due to various reasons, such as heavy rainfall, melting snow or ice, coastal storms, or the failure of dams or levees. Floods can range from minor inconveniences to devastating natural disasters (Fig. 13.1).

Fig. 13.1: Flood situation

13.1.1. Types of Floods

Floods can be classified based on their cause, speed of onset, and location. Some common types include:

- **River Floods (Fluvial Floods):** These occur when the water level in a river or stream rises and overflows its banks. They can develop slowly or rapidly depending on the duration and intensity of rainfall over the river's catchment area, snowmelt, and other factors.
- **Flash Floods (Pluvial Floods):** These are characterized by a rapid rise in water levels in a short amount of time (usually within 6 hours of heavy rainfall). They can be very dangerous due to their sudden onset and high water velocities. Flash floods often occur in small streams, urban areas with poor drainage, and mountainous regions.
- **Coastal Floods:** These occur when normally dry, low-lying land is inundated by seawater. They are often caused by storm surges associated with hurricanes or other coastal storms, high tides, and sea-level rise.
- **Urban Floods:** These happen in urbanized areas when rainfall exceeds the capacity of drainage systems.Impervious surfaces like roads and buildings prevent water from soaking into the ground, leading to surface runoff and flooding.
- **Groundwater Floods:** These occur when the level of water stored in the ground rises above the surface, often after prolonged periods of heavy rainfall. They tend to be localized and can last for extended periods.

13.1.2. Impact of Floods

The impacts of floods can be widespread and devastating:

- **Loss of Life and Injury:** Swift-moving floodwaters can easily sweep away people and cause drowning or other injuries.

- **Damage to Property:** Homes, businesses, infrastructure (roads, bridges, power lines), and agricultural lands can be severely damaged or destroyed by floodwaters and debris.
- **Displacement and Homelessness:** Floods can force people to evacuate their homes, leading to temporary or permanent displacement.
- **Contamination and Disease:** Floodwaters can become contaminated with sewage, chemicals, and other pollutants, leading to health hazards and the spread of waterborne diseases.
- **Economic Losses:** Floods can result in significant economic losses due to damage to property, infrastructure, and agriculture, as well as business disruptions.
- **Environmental Impacts:** Floods can cause soil erosion, siltation, damage to vegetation and wildlife habitats, and the spread of pollutants.

13.1.3. Flood Warnings

Flood warnings are crucial for mitigating the impacts of floods by allowing people time to prepare and evacuate if necessary. Warning systems typically involve:

- **Monitoring:** Continuous monitoring of rainfall, river levels, and coastal conditions.
- **Forecasting:** Using meteorological and hydrological models to predict when and where flooding is likely to occur.
- **Issuance of Warnings:** Communicating the flood risk to the public through various channels, such as radio, television, internet, and mobile alerts.

Flood warnings often have different levels to indicate the severity and likelihood of flooding, such as:

- **Flood Watch:** Conditions are favorable for flooding. Be prepared.
- **Flood Advisory:** Flooding is not expected to be severe, but could still be hazardous. Be aware.
- **Flood Warning:** Flooding is imminent or occurring. Take action.
- **Flash Flood Watch:** Conditions are favorable for flash flooding. Be prepared.
- **Flash Flood Warning:** A flash flood is imminent or occurring. Take immediate action.

13.1.4. Protective Measures in Agriculture

Protecting agriculture from floods is vital for food security and the livelihoods of farmers.Here are some detailed protective measures:

- **Planting Flood-Resilient Crop Varieties**
  - Opting for crop varieties known to tolerate waterlogging and temporary submergence can significantly reduce losses. Some varieties have adaptations like aerenchyma tissue that allows for oxygen transport in waterlogged conditions.
  - Diversifying crops grown can also help, as different crops have varying levels of flood tolerance.
- **Improved Drainage Systems**
  - **Surface Drainage:** Creating channels, ditches, and graded slopes to direct excess water away from fields.Regular maintenance to clear blockages is essential.
  - **Subsurface Drainage:** Installing underground pipes to remove excess water from the root zone, particularly in areas prone to waterlogging.
  - **Raised Beds and Ridges:** Planting crops on raised beds or ridges elevates them above potential waterlogging, especially beneficial for vegetables and high-value crops.
- **Construction of Protective Barriers**
  - **Bunds and Embankments:** Building raised barriers around fields, especially in low-lying areas, to prevent floodwater from inundating crops. The height and strength of these barriers should be appropriate for the expected flood levels.
- **Soil Management Practices**
  - **Increasing Organic Matter:** Soils rich in organic matter have better water infiltration and water-holding capacity, which can help manage excess water.
  - **No-Till Farming:** This practice can improve soil structure and reduce runoff.
  - **Cover Cropping:** Planting cover crops during off-seasons can protect the soil from erosion and improve its overall health.
- **Water Harvesting and Management**
  - Constructing farm ponds or reservoirs to collect and store excess rainwater during floods can help reduce immediate inundation and provide water for irrigation later.

- **Timely Harvesting**
  - If a flood is predicted, farmers should try to harvest mature crops as early as possible to minimize losses.
- **Crop Insurance**
  - Taking out crop insurance can provide financial protection against flood-related losses.
- **Weather Monitoring and Early Warning Systems**
  - Utilizing weather forecasts and flood warnings to take proactive measures, such as adjusting irrigation, moving livestock, and preparing for potential inundation.
- **Relocation of Vulnerable Infrastructure**
  - Locating farm buildings, storage facilities, and animal shelters on higher ground less prone to flooding.
- **Post-Flood Recovery**
  - Having plans for draining fields, removing debris, assessing soil damage, and addressing plant diseases or nutrient deficiencies after a flood event.

13.2. Droughts

A **drought** is a prolonged period of abnormally low rainfall, leading to a shortage of water. It's a creeping phenomenon, meaning it develops slowly and its onset and end can be difficult to pinpoint. Droughts can have significant impacts on agriculture, water resources, ecosystems, and the economy (Fig. 13.2). The definition of drought can vary depending on the region and the specific impacts being considered. Generally, it implies a significant deviation from normal precipitation patterns over an extended period.

Fig. 13.2: Droughts

13.2.1 Types of Droughts

Scientists have categorized droughts into several types to better understand their characteristics and impacts:

- **Meteorological Drought**
    - This type of drought is primarily based on the degree of dryness compared to a normal or average amount and the duration of the dry period.
    - It's usually defined by a significant deficit in precipitation over a specific period (e.g., a season, a year).
    - What constitutes a meteorological drought varies regionally because normal precipitation patterns differ greatly across the globe. For example, a few weeks without rain in a typically wet area might be considered a meteorological drought, while the same might be normal in an arid region.
- **Agricultural Drought**
    - This type of drought links meteorological or hydrological drought to agricultural impacts, focusing on factors like precipitation shortages, soil water deficits, and reduced groundwater or reservoir levels needed for irrigation.
    - It occurs when there isn't enough moisture in the soil to meet the needs of a particular crop at a particular time.
    - Agricultural drought considers the timing of rainfall deficits with the crop growth stages, the soil's water-holding capacity, and the specific water needs of different crops.
- **Hydrological Drought**
    - This type of drought refers to deficiencies in surface and subsurface water supplies.
    - It becomes evident in reduced streamflow, lower reservoir and lake levels, and declines in groundwater tables.
    - Hydrological droughts usually lag behind meteorological and agricultural droughts because it takes time for precipitation deficits to affect water bodies. Recovery from a hydrological drought can also be slow.
- **Socioeconomic Drought**
    - This type of drought associates the supply and demand of economic goods and services with elements of meteorological, agricultural, and hydrological drought.

- It occurs when the demand for water exceeds the supply due to a weather-related shortfall.
- Examples include water rationing, reduced hydropower production, and economic losses in drought-affected industries.

Sometimes, other types of droughts are also mentioned, such as

- **Ecological Drought:** This refers to a prolonged and widespread deficit in naturally available water that creates conditions that exceed the coping capacity of ecosystems, leading to impacts across species, communities, and ecosystem processes.
- **Flash Drought:** This is a rapidly intensifying drought characterized by unusually rapid rates of soil moisture depletion, typically occurring over a few weeks to a month. It's often associated with high temperatures, strong winds, and low humidity.
- **Snow Drought:** This occurs when there is an unusually low amount of snowpack, regardless of precipitation amounts. It can significantly impact water resources in regions that rely on snowmelt.

Understanding these different types of droughts helps in monitoring, assessing impacts, and developing appropriate response strategies.

13.2.2. Meteorological Drought

The India Meteorological Department (IMD) classifies meteorological droughts based on the **rainfall deficiency** from the long-term average or normal rainfall for a given area. Here's how they typically categorize it:

- **No Drought:** Rainfall deviation is within +19% to -19% of the normal.
- **Mild Drought:** Rainfall deficiency is between -20% to -39% of the normal.
- **Moderate Drought:** Rainfall deficiency is between -40% to -59% of the normal.
- **Severe Drought:** Rainfall deficiency is greater than -60% of the normal.

Additionally, IMD considers a year as an **All-India Drought Year** if the rainfall deficiency for the country as a whole exceeds 10%, and if more than 20-40% of the total area of the country is affected by drought conditions.

It's important to note that these are broad classifications, and the specific criteria and thresholds can sometimes be refined or vary slightly depending on the region and the purpose of the analysis.

13.2.3. Hydrological Drought

In India, hydrological drought, characterized by deficits in surface and subsurface water supplies, is monitored using several indicators

- **River/Stream Flow:** The Central Water Commission (CWC) monitors the water levels and flow rates in major rivers and their tributaries across the country. Significant deviations from normal flow for an extended period can indicate a hydrological drought. The **Standardized Streamflow Index (SSI)** is also used to quantify streamflow drought.
- **Reservoir Levels:** The levels of water in major reservoirs are tracked. A substantial and prolonged decline in reservoir storage compared to historical averages signals a potential hydrological drought. The **Reservoir Storage Index (RSI)** is used for this purpose, often categorizing deficits as normal, mild, moderate, severe, or extreme based on the percentage deviation from the mean storage.
- **Groundwater Levels:** Monitoring the depth to the water table in wells provides insights into groundwater storage. A persistent decline below normal levels indicates a groundwater drought, which is a component of hydrological drought. The **Groundwater Drought Index (GWDI)** is used to assess this.
- **Lakes and Other Water Bodies:** The water levels in significant lakes and other surface water bodies are also observed.
- **Key Institutions Involved in Monitoring**
- **Central Water Commission (CWC):** Primarily responsible for monitoring surface water resources (rivers and reservoirs).
- **India Meteorological Department (IMD):** While primarily focused on meteorological drought through rainfall monitoring and indices like the Aridity Anomaly Index (AAI) and Standardized Precipitation Index (SPI), prolonged meteorological drought is a precursor to hydrological drought.
- **Central Ground Water Board (CGWB):** Monitors groundwater levels across the country.
- **Declaration of Hydrological Drought in India**

There isn't a single, nationwide, formal declaration of "hydrological drought" made by a central authority in the same way that meteorological drought is assessed by the IMD based on rainfall deficits. The declaration of drought, including aspects related to hydrological conditions, is primarily the **prerogative of the State Governments**. They consider various factors, including:

- **Rainfall deficiency (meteorological drought as indicated by IMD).**
- **Status of water levels in reservoirs and rivers.**
- **Groundwater availability.**
- **Impacts on agriculture and other sectors.**

Based on these assessments, State Governments declare certain areas as drought-affected and implement relief measures.

However, at the national level, the monitoring data from CWC and CGWB helps in understanding the overall hydrological stress in different regions. This information is crucial for:

- **Advising state governments.**
- **Planning for water resource management.**
- **Allocating resources.**

In summary: Hydrological drought in India is monitored through the observation of river flows, reservoir and lake levels, and groundwater tables by agencies like CWC and CGWB. While the IMD monitors meteorological drought which often leads to hydrological drought, the formal declaration of drought, considering hydrological factors alongside others, is primarily done by the respective State Governments.

13.2.4. Agricultural Droughts

Agricultural drought occurs when the amount of moisture in the soil no longer meets the water demands of a particular crop during its growth stages, leading to reduced growth and yield. It links meteorological and hydrological droughts to their impact on agriculture. It's not just about the amount of rainfall, but also about the timing of rainfall, soil water holding capacity, and the specific needs of the crops.

- **Types of Agricultural Droughts:** The types of agricultural droughts, including permanent drought are:
  - **Early Season Drought:** This occurs due to a delay in the onset of the monsoon or prolonged dry spells after early sowing. It can hinder germination and early vegetative growth, leading to poor crop establishment.
  - **Mid-Season Drought:** This happens during the vegetative or flowering stages of crop growth due to long gaps between successive rainfall events, when stored soil moisture becomes insufficient to meet the crop's transpiration demands. This can lead to stunted growth, reduced leaf area, and ultimately lower yields.
  - **Late Season Drought:** This occurs due to the early cessation of rainfall, causing water stress during the grain-filling or maturity stages of the crop. This can result in shriveled grains and reduced overall yield and quality.

- **Invisible Drought:** This can occur even with some rainfall, where the daily rainfall is insufficient to meet the daily water needs of the plants, leading to sub-optimal growth and yield reduction without obvious signs of severe wilting.
- **Apparent Drought:** This arises from a mismatch between the cropping pattern and the available moisture. For example, sufficient rainfall for one crop might be insufficient for another with higher water requirements.
- **Physiological Drought:** In this case, water is available in the soil, but plants cannot absorb it due to physiological reasons, such as high osmotic pressure in saline or alkaline soils.
- **Soil Drought:** This broadly refers to a deficiency of water in the soil, which can be physical (actual shortage) or physiological (water unavailable to plants).
- **Permanent Drought:** This refers to arid or hyper-arid climates where rainfall is consistently very low, making it extremely challenging or impossible to sustain rain-fed agriculture without significant irrigation. In these regions, drought conditions are the norm rather than an anomaly.

The impact of agricultural drought depends on the **timing**, **duration**, and **intensity** of the water stress relative to the crop's growth stage. Certain growth stages, like flowering and grain filling, are often more sensitive to water deficits.

13.2.5. Impact of Agricultural Droughts

The impacts of agricultural droughts are far-reaching and can be devastating, affecting not only crop production but also the broader economy, environment, and society. Here are some key impacts:

- **Reduced Crop Yields and Failures**
    - The most direct impact is the significant reduction in crop yields or complete crop failure due to insufficient water at critical growth stages. This leads to lower overall agricultural output.
- **Livestock Losses**
    - Droughts can lead to a scarcity of pasture and drinking water for livestock, resulting in reduced productivity, poor health, and even death of animals. Farmers may be forced to sell their livestock prematurely, often at reduced prices.

- **Economic Losses**
  - **Farmers:** Reduced income due to lower yields and livestock losses.Increased costs for irrigation (if available), fodder, and other resources.
  - **Agricultural Sector:** Decline in the overall value of agricultural production, affecting related industries like food processing, transportation, and agricultural input suppliers.
  - **Consumers:** Potential increase in food prices due to shortages.
  - **Regional and National Economy:** Reduced GDP growth, particularly in economies heavily reliant on agriculture.
- **Food Insecurity**
  - Widespread crop failures can lead to food shortages, impacting food availability and accessibility, especially for vulnerable populations. This can exacerbate hunger and malnutrition.
- **Water Scarcity**
  - Agricultural drought exacerbates water scarcity for other sectors as well, as irrigation demands increase while overall water availability decreases. This can lead to conflicts over water resources.
- **Soil Degradation**
  - Prolonged dry periods can make the soil more susceptible to wind and water erosion once rainfall returns. Loss of vegetation cover during drought also contributes to soil degradation.
- **Increased Pests and Diseases**
  - Ironically, some pests and diseases can become more prevalent during or after a drought, further impacting crop and livestock health.
- **Social Impacts**
  - **Rural Livelihoods:** Droughts can severely impact the livelihoods of farmers and agricultural laborers, leading to unemployment, migration, and social unrest in extreme cases.
  - **Health:** Malnutrition and stress-related health issues can increase in drought-affected communities.
- **Environmental Impacts**
  - Loss of biodiversity due to habitat degradation.
  - Increased risk of wildfires in dry vegetation.
  - Dust storms due to dry and exposed soil.

In summary, agricultural droughts have a cascading effect, starting with water deficits in the soil and leading to significant economic, social, and environmental consequences. Understanding these impacts is crucial for developing effective drought preparedness and mitigation strategies.

13.3. Preparedness and Strategies for Mitigating Agricultural Droughts

Effective management of agricultural drought involves a combination of proactive **preparedness strategies**implemented before a drought and responsive **mitigation strategies** applied during or immediately after a drought. The goal is to build resilience in agricultural systems and minimize the adverse impacts of water scarcity.

13.3.1. Preparedness Strategies

These measures aim to reduce vulnerability to future droughts:

- **Choice of Drought-Resistant Crop Varieties**
  - Selecting and promoting the cultivation of crop varieties that are inherently more tolerant to water stress. This includes varieties with deeper root systems, efficient water use mechanisms, and the ability to withstand prolonged dry spells (e.g., certain types of millets like Bajra and Sorghum, drought-tolerant pulses like Pigeon Pea and Chickpea, and specific varieties of maize and cotton).
- **Adjustment of Sowing Periods**
  - Modifying planting schedules to avoid the most drought-vulnerable periods of the growing season. This might involve early sowing to allow crops to mature before anticipated dry spells, or delaying sowing until sufficient soil moisture is available. Utilizing seasonal weather forecasts is crucial for making informed decisions about optimal sowing times.
- **Water-Efficient Irrigation Techniques**
  - Implementing advanced irrigation methods like drip and sprinkler systems to deliver water directly to plants, minimizing losses from evaporation and runoff.
  - Utilizing soil moisture monitoring to optimize irrigation scheduling and avoid unnecessary water use.
- **Soil Moisture Conservation Practices**
  - Employing techniques such as mulching, conservation tillage, contour farming, terracing, and increasing soil organic matter to enhance the soil's ability to retain water and reduce evaporation.

- **Water Harvesting and Storage**
    - Constructing farm ponds, check dams, and other structures to capture and store rainwater for supplemental irrigation during dry periods.
- **Crop Diversification**
    - Growing a variety of crops with different water needs and drought tolerances to reduce the risk of total loss.
- **Agroforestry**
    - Integrating trees into farming systems to improve soil health, microclimate, and overall resilience.
- **Weather Monitoring and Early Warning Systems:**
    - Utilizing and disseminating weather forecasts and drought advisories to enable timely decision-making by farmers.
- **Crop Insurance**
    - Providing financial protection to farmers against crop losses due to drought.
- **Community-Based Preparedness**
    - Fostering collaboration and knowledge sharing among farmers to promote the adoption of drought-resilient practices.

13.3.2. Mitigation Strategies

These actions are taken when drought conditions are present:

- **Prioritized Water Allocation**
    - Strategically distributing limited water resources to the most critical crops or growth stages.
- **Supplemental Irrigation**
    - Using available stored water or groundwater to provide essential irrigation.
- **Adjusted Agronomic Practices**
    - Modifying fertilization, pest and disease management, and plant density to suit drought conditions.
- **Livestock Support**
    - Providing alternative feed and water sources for livestock, and managing herd sizes.
- **Strict Water Conservation**
    - Implementing all possible water-saving measures on the farm.

- **Post-Drought Recovery Support**
  - Assisting farmers with soil rehabilitation, provision of seeds, and other resources for recovery.

By integrating the selection of appropriate crop varieties and the timing of sowing with other preparedness and mitigation measures, agricultural systems can become significantly more resilient to the challenges posed by drought.

13.4. Hail Storms

Hailstorms are weather events where precipitation falls in the form of **hailstones**, which are balls or irregular lumps of ice. Hail forms in strong thunderstorm clouds, particularly those with intense updrafts that carry water droplets into very cold regions of the atmosphere where they freeze (Fig. 13.3). These ice particles can grow as they collide with supercooled water droplets, which then freeze onto the hailstone. Hailstones can vary in size from small pebbles to as large as softballs or even larger.

Fig. 13.3: Hail storms

13.4.1. Impact of Hailstorms on Agriculture

Hailstorms, even if short-lived, can cause significant and widespread damage to agricultural crops:

- **Physical Damage to Plants:** Hailstones can shred leaves, break stems, bruise fruits and vegetables, and even uproot young plants. This direct physical damage reduces the plant's ability to photosynthesize, weakens its structure, and makes it more susceptible to diseases and pests.
- **Damage to Produce:** Fruits, grains, and vegetables can be directly damaged, making them unmarketable. Bruising can lead to spoilage and reduced shelf life.
- **Delayed Growth and Reduced Yields:** Plants damaged by hail often struggle to recover, diverting energy to repair rather than growth and production, leading to delayed maturity and lower overall yields.

- **Soil Erosion:** Intense hailstorms are often accompanied by heavy rainfall, which, combined with damaged or removed plant cover, can lead to increased soil erosion and loss of fertile topsoil.
- **Post-Harvest Losses:** Crops that are near maturity or already harvested but still in the fields can also be damaged by hail.

The severity of the impact depends on the size and density of the hailstones, the duration of the hailstorm, the growth stage of the crops, and the type of crop. Some crops are more vulnerable to hail damage than others.

13.4.2. Hailstorm Warning Systems

Timely warnings can allow farmers to take some protective measures. Hailstorm warning systems rely on a combination of:

- **Weather Radar:** Doppler radar can detect the presence of hail within thunderstorms by identifying areas of high reflectivity.Dual-polarization radar can further differentiate between rain, hail, and ice pellets and even estimate hail size.
- **Satellite Imagery:** While not as direct as radar for hail detection, satellite data can help identify severe thunderstorms that have the potential to produce hail.
- **Surface Observations:** Reports from weather stations and trained spotters can confirm the occurrence and size of hail.
- **Numerical Weather Prediction Models:** These models can forecast the development and movement of severe thunderstorms that may produce hail.
- **Specialized Hail Detection Algorithms:** Systems like the National Severe Storms Laboratory's (NSSL) Hail Detection Algorithm (HDA) and Multi-Radar, Multi-Sensor (MRMS) Maximum Estimated Hail Size (MESH) provide estimates of hail probability and size.
- **Commercial Services:** Some private companies offer specialized hail forecasting and alert services, often with higher resolution and lead times, using proprietary algorithms and data integration (e.g., HailSensNow, HailSens360). These may provide alerts via SMS, email, or apps.

While forecasting the exact location and timing of hailstorms remains challenging, these systems provide valuable information for risk areas.

13.4.3. Protective Measures in Agriculture Against Hailstorms

Farmers can employ several strategies to protect their crops from hail damage:

- **Anti-Hail Nets**
    - These are mesh nets installed over crops (especially high-value ones like fruits, vegetables, and vineyards) using a support structure.
    - The nets intercept hailstones, reducing their impact and preventing direct contact with the plants and produce.
    - While effective, the installation can be costly, making it more suitable for high-value crops or smaller areas.
- **Protective Covers**
    - For smaller plants or nursery stock, temporary covers can be used when a hailstorm is imminent.
- **Hail-Resistant Crop Varieties**
    - Some crop varieties may have structural characteristics that make them slightly more resistant to hail damage (e.g., stronger stems, different leaf orientation). However, complete resistance is rare.
- **Crop Insurance**
    - While not a direct protective measure, crop insurance can provide financial compensation to farmers who experience losses due to hail, helping them recover.
- **Promoting Healthy Soil**
    - Healthy soils with good organic matter can help plants recover better from stress, including hail damage.
- **Pod Sealants**
    - In some crops like rapeseed, polymer and surfactant blends can be sprayed onto pods to make them more resistant to shattering from hail.
- **Integrated Pest and Disease Management**
    - Hail damage can create entry points for pests and diseases. Prompt management after a hailstorm is crucial to prevent secondary damage.
- **Post-Hail Management**
    - Removing severely damaged plant parts to encourage regrowth.
    - Applying appropriate fertilizers to aid recovery.
    - Implementing phytosanitary treatments to prevent infections.

While completely preventing hail damage is not usually possible, these protective measures can significantly reduce the extent of the losses. The choice of method often depends on the type of crop, the scale of farming, and the economic feasibility.

13.5. Frost

Frost is the deposition of ice crystals on a surface when the temperature of that surface falls below the freezing point of water (0°C or 32°F). It occurs through a process called deposition, where water vapor in the air changes directly into ice without first becoming liquid water (Fig. 13.4).

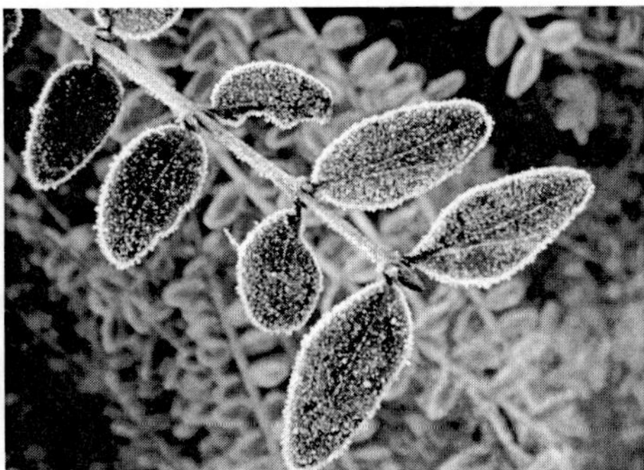

Fig. 13.4: Frost deposition

13.5.1. When Does Frost Occur?

Frost typically forms on clear, calm nights when

- The air temperature near the ground falls to or below freezing.
- The surface temperature is even lower than the air temperature due to radiative cooling (the ground loses heat to the atmosphere).
- There is sufficient moisture in the air.

13.5.2. Impact of Frost on Crops

Frost can severely damage crops, especially during vulnerable growth stages:

- **Cell Damage:** Ice crystals form within plant tissues, disrupting cell structure and causing dehydration.
- **Wilting and Death:** Damaged cells can lead to wilting, browning, and ultimately the death of plant parts or the entire plant.
- **Reduced Yields:** Frost can damage flowers, fruits, and developing seeds, significantly reducing crop yields.
- **Increased Susceptibility to Diseases:** Frost-damaged plants are more vulnerable to fungal and bacterial infections.
- **Delayed Maturity:** Frost can delay the maturity of crops, making them more susceptible to other weather hazards.

The severity of frost damage depends on the temperature, duration of the frost, the growth stage of the crop, and the plant species.

13.5.3. Frost Warnings

Frost warnings are issued by meteorological agencies to alert farmers and the public about the risk of frost. These warnings typically include:

- **Predicted Minimum Temperatures:** Indicating how low the temperature is expected to fall.
- **Duration of the Frost:** How long the below-freezing temperatures are expected to last.
- **Areas at Risk:** Specifying the regions where frost is most likely.
- **Timing of the Frost:** When the frost is expected to occur (e.g., overnight).

13.5.4. Frost Protection Measures in Agriculture

Farmers use various methods to protect their crops from frost:

- **Covering Plants**
  - Using blankets, plastic sheets, or row covers to trap heat radiating from the ground.
  - Effective for small plants and high-value crops.
- **Irrigation**
  - Sprinkling water on plants before and during a frost can protect them. As the water freezes, it releases heat, keeping the plant tissues near 0°C.
- **Wind Machines**
  - Large fans that mix warmer air from aloft with cooler air near the ground, preventing the formation of a strong temperature inversion.
- **Heaters**
  - Using heaters or smudge pots in orchards or vineyards to raise the air temperature.
- **Site Selection**
  - Avoiding planting frost-sensitive crops in low-lying areas where cold air tends to accumulate.
- **Delayed Planting**
  - Planting crops later in the season to avoid the highest risk of frost.

- **Crop Selection**
  - Choosing crop varieties that are more frost-tolerant.
- **Pruning**
  - Proper pruning can improve air circulation and reduce frost damage.
- **Promoting Healthy Soil**
  - Healthy soils retain more heat and can help plants withstand frost better.
- **Protective Sprays**
  - Some chemical sprays can offer limited protection by reducing ice nucleation on plant surfaces.

The most effective frost protection strategy depends on the type of crop, the severity of the expected frost, and the economic feasibility.

13.6. Heat Waves

Heat waves are prolonged periods of abnormally hot weather that can have severe impacts on human health, agriculture, and the environment. In India, the India Meteorological Department (IMD) defines heat waves based on specific temperature thresholds. The IMD defines heat waves as follows:

- **Plain Areas**: A maximum temperature of 40°C (104°F) or more is considered a heat wave.
- **Coastal Areas**: A maximum temperature of 37°C (98.6°F) or more is considered a heat wave.
- **Hilly Areas**: A maximum temperature of 30°C (86°F) or more is considered a heat wave.
- **Departure from Normal Temperature**
  - **Heat Wave**: A departure of 4.5°C (8.1°F) to 6.4°C (11.5°F) above normal.
  - **Severe Heat Wave**: A departure of more than 6.4°C (11.5°F) above normal.
- **Actual Maximum Temperature**
  - **Heat Wave**: 45°C (113°F) or more.
  - **Severe Heat Wave**: 47°C (116.6°F) or more.

13.6.1 Impact on Crops

Heat waves can have devastating effects on crops, including:

- **Reduced Yields**: High temperatures can cause crops to wilt, reducing yields and affecting productivity.

- **Crop Failure**: Extreme heat can damage or destroy crops, leading to complete crop failure.
- **Lower Quality Produce**: Heat stress can reduce the quality of fruits and vegetables, making them less marketable.
- **Increased Water Demand**: Heat waves can strain irrigation systems, depleting groundwater reserves and affecting crop growth.

13.6.2. Impact on Livestock

Heat waves can also impact livestock, causing:

- **Heat Stress**: High temperatures can lead to heat exhaustion and death in animals.
- **Reduced Milk Production**: Heat stress can reduce milk production and affect meat quality.
- **Increased Mortality**: Extreme heat can lead to increased mortality rates in livestock.

13.6.3. Heat Wave Warnings

The IMD issues heat wave warnings based on the criteria mentioned above (Fig. 13.5). These warnings are crucial for taking proactive measures to protect crops, livestock, and human health.

Fig. 13.5: Heat wave conditions

13.6.4. Protective Measures

To mitigate the effects of heat waves, farmers can take the following measures:

- **Irrigation Management**: Irrigate crops early and often to prevent moisture stress.
- **Shading**: Use shade cloth or other methods to provide relief from direct sunlight.
- **Heat-Tolerant Crops**: Plant heat-tolerant crop varieties to reduce the impact of heat waves.

- **Soil Management**: Use mulch or covers to reduce soil evaporation and retain moisture.
- **Weather Forecasting**: Monitor weather forecasts to prepare for heat waves and take proactive measures.

13.6.5. Government Initiatives

The government can take several steps to address heat waves, including:

- **Improving Forecasting**: Enhance heat wave forecasting and warning systems to provide early warnings.
- **Public Awareness**: Raise public awareness about heat wave risks and precautions to protect vulnerable populations.
- **Water Management**: Implement water conservation measures to reduce the impact of heat waves on water resources.
- **Support to Farmers**: Provide support to farmers, including irrigation subsidies, crop insurance programs, and other measures to reduce the impact of heat waves on agriculture.

By understanding heat waves and taking proactive measures, we can reduce the impact of these extreme weather events on agriculture, livestock, and human health.

13.7. Cold Waves

Cold waves are prolonged periods of abnormally cold weather that can have significant impacts on human health, agriculture, and the environment. In India, the India Meteorological Department (IMD) defines cold waves based on specific temperature thresholds (Fig. 13.6).

Fig. 13.6: Cold wave situation

13.7.1. Criteria for Cold Waves (IMD)

The IMD defines cold waves as follows:

- **Plain Areas**: A minimum temperature of 10°C (50°F) or less is considered a cold wave.

- **Hilly Areas**: A minimum temperature of 0°C (32°F) or less is considered a cold wave.
- **Departure from Normal Temperature**
    - **Cold Wave**: A departure of 4.5°C (8.1°F) to 6.4°C (11.5°F) below normal.
    - **Severe Cold Wave**: A departure of more than 6.4°C (11.5°F) below normal.

13.7.2. Impact on Crops

Cold waves can have significant impacts on crops, including

- **Frost Damage**: Frost can damage or destroy crops, leading to significant losses.
- **Reduced Yields**: Cold temperatures can reduce crop growth and yields.
- **Crop Failure**: Extreme cold can damage or destroy crops, leading to complete crop failure.
- **Delayed Planting**: Cold waves can delay planting and affect crop development.

13.7.3. Impact on Livestock

Cold waves can also impact livestock, causing:

- **Hypothermia**: Prolonged exposure to cold temperatures can lead to hypothermia in animals.
- **Respiratory Issues**: Cold stress can exacerbate respiratory issues in animals.
- **Reduced Milk Production**: Cold stress can reduce milk production and affect meat quality.
- **Increased Mortality**: Extreme cold can lead to increased mortality rates in livestock.

13.7.4. Cold Wave Warnings

The IMD issues cold wave warnings based on the criteria mentioned above. These warnings are crucial for taking proactive measures to protect crops, livestock, and human health.

13.7.5. Protective Measures

To mitigate the effects of cold waves, farmers can take the following measures:

- **Provide Shelter**: Provide shelter for animals to protect them from cold winds and extreme temperatures.
- **Use Bedding**: Use dry bedding to keep animals warm and dry.

- **Increase Nutrition**: Increase nutrition for animals to help them cope with cold stress.
- **Protect Crops**: Use techniques like mulching, covering, or using windbreaks to protect crops from frost and cold winds.
- **Monitor Weather**: Monitor weather forecasts to prepare for cold waves and take proactive measures.

13.7.6. Government Initiatives

The government can take several steps to address cold waves, including:

- **Improving Forecasting**: Enhance cold wave forecasting and warning systems to provide early warnings.
- **Public Awareness**: Raise public awareness about cold wave risks and precautions to protect vulnerable populations.
- **Support to Farmers**: Provide support to farmers, including subsidies for crop insurance, irrigation, and other measures to reduce the impact of cold waves on agriculture.
- **Animal Husbandry Support**: Provide support to animal husbandry sector, including measures to protect livestock from cold stress.

By understanding cold waves and taking proactive measures, we can reduce the impact of these extreme weather events on agriculture, livestock, and human health.

13.8. Strong and Gusty Winds

Strong winds are characterized by their speed and force, capable of causing physical damage. Gusty winds are sudden, brief increases in wind speed (Fig. 13.7). Both can pose significant challenges to agriculture.

Fig. 13.7: Gusty winds

13.8.1. Wind Speeds Damaging Agriculture

Damage to agriculture typically starts becoming significant at wind speeds above **30 miles per hour (around 48 km/h)**. The impact depends on the duration, gustiness, and the state of the plants.

- **Mechanical Damage**
  - Shredding of leaves.
  - Tearing off flowers and fruits.
  - Breaking branches and stems.
  - Uprooting of shallow-rooted plants and trees.
  - Lodging (bending or flattening) of cereal crops.
- **Physiological Impacts**
  - Increased transpiration leading to rapid drying of plants, potentially causing wilting and browning, especially if roots cannot keep up with water loss.
  - Exacerbation of stress from freezing conditions, extreme heat, and drought.
  - Disruption of pollination by affecting insect activity or directly damaging flowers.
- **Soil Erosion:** Strong winds, especially in dry and uncovered areas, can cause significant soil erosion, leading to loss of fertile topsoil and damage to young seedlings through abrasion from wind-borne particles.
- **Damage to Infrastructure:** Damage to greenhouses, plastic tunnels, and other farm structures.

13.8.2. Impact of Gale

A **gale** is defined as a strong wind, typically with sustained speeds ranging from **39 to 54 miles per hour (63 to 87 km/h)**, which corresponds to force 8 and sometimes 9 on the Beaufort scale.

- **Widespread and Severe Mechanical Damage:** Increased likelihood of significant structural damage to plants, including breaking of large branches and uprooting of even relatively established plants.
- **Severe Soil Erosion:** Especially in open fields, gales can remove substantial amounts of topsoil.
- **Damage to Farm Buildings:** Higher risk of damage to sheds, barns, and other agricultural infrastructure.
- **Disruption of Farm Activities:** Makes outdoor work hazardous and can halt essential farming operations.

13.8.3. Warning Systems for Strong Winds

Warning systems for strong winds relevant to agriculture often come as part of general weather forecasts and severe weather alerts issued by meteorological departments like the IMD. These may include:

- **Wind Speed Forecasts:** Predicting sustained wind speeds and potential gusts.
- **Gale Warnings:** Specific alerts when gale-force winds are expected, particularly important in coastal agricultural areas.
- **Impact Advisories:** Sometimes, warnings will mention potential impacts on agriculture.
- **Local Weather Stations:** Farmers using on-site weather stations can get real-time wind speed data.

13.8.4. Protective Measures Against Strong Winds

Farmers can implement several protective measures:

- **Windbreaks**
  - Planting rows of trees, shrubs, or using artificial barriers like fences to reduce wind speed in fields. Effective windbreaks can reduce wind velocity significantly for a distance several times their height (Fig. 13.8).
  - Choose windbreak species that are dense or multi-stemmed to the ground.
- **Shelterbelts**
  - Similar to windbreaks but often more extensive, using multiple rows of trees and shrubs (Fig. 13.8).

Fig. 13.8: Windbreaks and shelterbelts

Staking and Support

- Providing physical support to vulnerable plants, especially tall or fruiting crops, to prevent lodging or breakage.

- **Pruning**
  - Pruning trees in windy areas to reduce the density of the crown, allowing wind to pass through more easily.
- **Soil Management**
  - Maintaining adequate soil moisture to help plants withstand wind stress.
  - Using cover crops and maintaining crop residues to protect the soil from wind erosion.
- **Netting and Covers**
  - Using netting or temporary covers for high-value crops to provide some protection from wind damage.
- **Site Selection**
  - Choosing planting sites that are naturally more sheltered from prevailing winds.
- **Crop Selection**
  - Opting for crop varieties that are known to have better wind tolerance (e.g., those with stronger stems or more flexible structures).

The specific measures will depend on the local climate, the types of crops grown, and the frequency and intensity of strong winds in the area.

14

Microclimatic Modifications

Modifying microclimates in agriculture involves altering the local climate conditions surrounding crops or plants to create a more favorable environment for growth. This can be achieved through various techniques when the crops are subjected to weather related stresses.

14.1. To Mitigate Moisture Stress during Droughts

When the crops are under moisture stress, the principles involved are conservation of soil moisture and reduce the evapotranspiration from crops. The moisture conservation can be achieved by suppressing soil evaporation using mulches (Fig. 14.1). The evapotranspiration can be reduced using anti transpirants. These are chemicals sprayed on crops. Some antitranspirents close the stomata of the leaves to reduce evapotranspiration. Some chemicals increase the reflectance of radiation from the crops, thereby reducing heat load on crops. When crops are grown under irrigation in isolated patches, micro shelter belts can be provided across the predominant wind direction to reduce the increased evapotranspiration from advection. Advection is phenomena of horizontal transport of heat due to strong hot and dry winds.

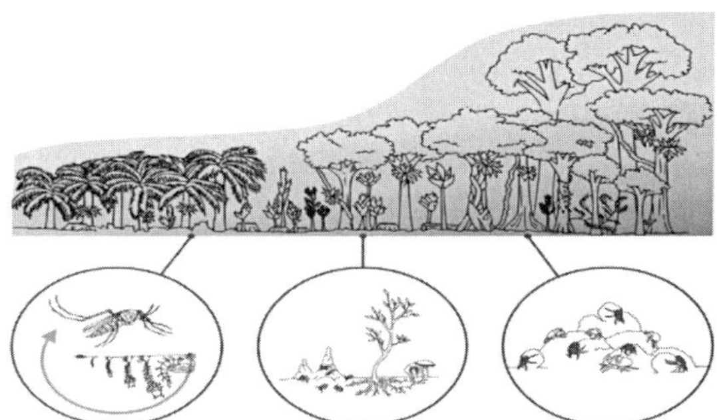

Fig. 14.1: Microclimate modification

14.2. To Suppress Damage due to Hail Storms

To suppress damage to crops due to hail storms using microclimate modification, several strategies can be employed:

- **Hail nets**: Installing hail nets or screens over crops can protect them from direct hail impact. These nets can be particularly effective for high-value crops like fruits and vegetables.
- **Windbreaks**: While windbreaks can't prevent hail, they can reduce wind speed and potentially minimize some hail damage by altering the microclimate.
- **Greenhouses**: Growing crops in greenhouses can provide a physical barrier against hail. Greenhouses offer a controlled environment that shields crops from extreme weather conditions.

14.3. To Suppress Impact of Heat Waves

- Microclimate modification can help suppress heat wave effects on crops by employing techniques such as providing shade through tree canopies or artificial shading structures, using mulch or cover crops to retain soil moisture and reduce soil temperature, and implementing efficient irrigation systems like drip irrigation. Additionally, windbreaks can reduce wind speed and minimize moisture loss, while greenhouses or protective coverings can offer a controlled environment that shields crops from extreme temperatures. By modifying the microclimate, farmers can reduce heat stress on crops, promote healthy growth, and improve yields even during heat waves.

14.4. To Protect Crops from Frost

- Microclimate modification can play a crucial role in preventing damage to crops due to frost by employing various techniques that alter the local climate conditions surrounding the crops. One effective method is the use of wind machines or wind turbines that circulate warmer air above the crop canopy down to the ground level, thereby preventing the formation of frost. Another technique is the application of sprinkler irrigation, where water is sprayed over the crops; as the water freezes, it releases heat and protects the plants from frost damage. Additionally, the use of row covers or other protective coverings can trap warm air close to the soil and protect the crops from frost. Furthermore, creating a microclimate through the strategic placement of windbreaks or shelterbelts can also help in reducing the risk of frost damage by minimizing cold air drainage and preventing harsh winds from directly impacting the crops. Orchards and vineyards often use heaters or smudge pots to warm the surrounding air during frost events, which can be an effective, though sometimes costly, method of frost protection. By modifying the microclimate in these ways, farmers can significantly reduce the risk of frost damage to their crops, ensuring healthier plants and better yields even in frost-prone areas.

14.5. To Protect Crops from Strong Winds

- Protection of crops from strong winds can be achieved through various microclimate modification techniques. One effective method is the establishment of windbreaks or shelterbelts, which are rows of trees, shrubs, or other vegetation planted perpendicular to the prevailing wind direction. These barriers help reduce wind speed, thereby minimizing damage to crops, preventing soil erosion, and reducing moisture loss through evapotranspiration. Windbreaks can also alter the microclimate by creating a more favorable environment for crop growth, leading to improved yields and better plant health. Additionally, the use of artificial windbreaks such as burlap screens, snow fencing, or plastic mesh screens can provide temporary or permanent protection for crops. These screens can be particularly useful for high-value crops or in areas where wind damage is frequent and severe. Furthermore, the strategic planting of crops in areas naturally shielded from strong winds, such as valleys or lee slopes, can also mitigate wind damage. By modifying the microclimate through these techniques, farmers can protect their crops from wind-induced stress, breakage, and loss, ultimately enhancing agricultural productivity and sustainability

15

Climate Normals for Crops and Livestock

Climate normals for crops and livestock refer to the long-term average weather conditions, such as temperature, precipitation, and humidity that are suitable for specific crops and livestock in a particular region. These normals help farmers and agricultural planners understand the typical climate conditions that support optimal growth, productivity, and health of crops and livestock. By knowing the climate normals, farmers can make informed decisions about:

- Crop selection and breeding
- Planting and harvesting schedules
- Livestock management and breeding
- Irrigation and water management
- Pest and disease management

15.1. Crop Weather Calendars

A Crop Weather Calendar (CWC) is a valuable agrometeorological tool used to align the crop growth stages with prevailing weather conditions to optimize agricultural practices. It helps farmers, extension workers, and planners make informed decisions by integrating weather information with the crop‘s physiological needs (Fig. 15.1).

The crop-weather calendars are used in agromet-advisory services. With the week as the basic time unit, the compiled information can assist forecasters in framing weather warnings and forecasts directed at farmers. Crop-weather calendars provide information on crop growth stages, normal weather for crop growth, warnings to be issued based on prevailing weather condition, water requirement of crops during their various phenophases and weather conditions favourable for development of crop pests and diseases. These calendars are useful for crop planning, irrigation scheduling and plant protection measures.

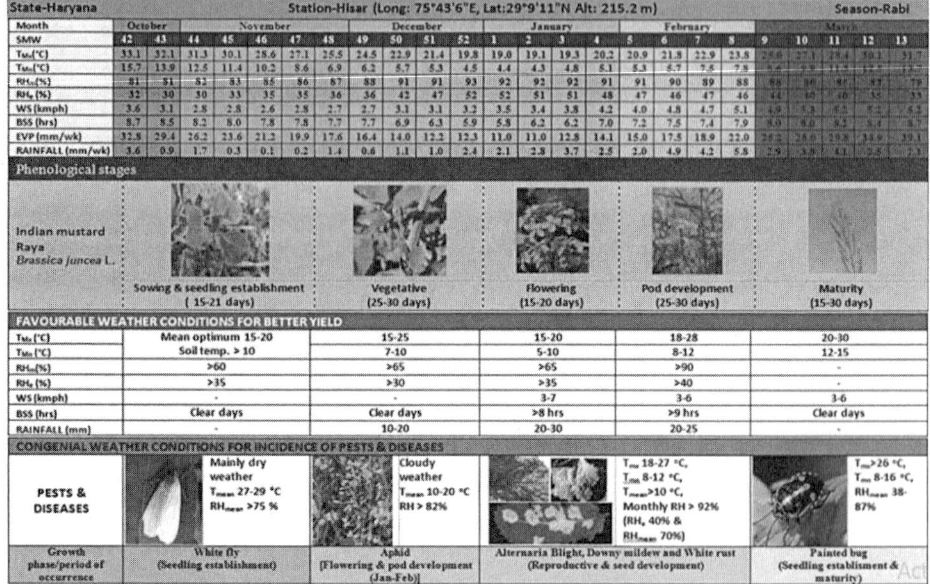

State-Haryana	Station-Hisar (Long: 75°43'6"E, Lat:29°9'11"N Alt: 215.2 m)																					Season-Rabi		
Month	October		November					December				January				February				March				
SMW	42	43	44	45	46	47	48	49	50	51	52	1	2	3	4	5	6	7	8	9	10	11	12	13
T_{Max}(°C)	33.1	32.1	31.3	30.1	28.6	27.1	25.5	24.5	22.9	21.4	19.8	19.0	19.1	19.3	20.2	20.9	21.8	22.9	23.8	[illegible]	27.1	28.4	30.1	31.7
T_{Min}(°C)	15.7	13.9	12.5	11.4	10.2	8.6	6.9	6.2	5.7	5.3	4.5	4.4	4.3	4.8	5.1	5.3	5.7	7.5	7.8	[illegible]	[illegible]	[illegible]	[illegible]	[illegible]
RH_{M}(%)	81	81	82	83	85	86	87	88	91	91	93	92	92	92	91	91	90	89	88	[illegible]	[illegible]	[illegible]	[illegible]	79
RH_{E} (%)	32	30	30	33	35	35	36	36	42	47	52	52	51	51	48	47	46	47	46	44	[illegible]	40	[illegible]	33
WS (kmph)	3.6	3.1	2.8	2.8	2.6	2.8	2.7	2.7	3.1	3.1	3.2	3.5	3.4	3.8	4.2	4.0	4.8	4.7	5.1	4.9	[illegible]	[illegible]	[illegible]	[illegible]
BSS (hrs)	8.7	8.5	8.2	8.0	7.8	7.8	7.7	7.7	6.9	6.3	5.9	5.8	6.2	6.2	7.0	7.2	7.5	7.4	7.9	[illegible]	[illegible]	[illegible]	[illegible]	[illegible]
EVP (mm/wk)	32.8	29.4	26.2	23.6	21.2	19.9	17.6	16.4	14.0	12.2	12.3	11.0	11.0	12.8	14.1	15.0	17.5	18.9	22.0	[illegible]	[illegible]	29.8	[illegible]	39.1
RAINFALL (mm/wk)	3.6	0.9	1.7	0.3	0.1	0.2	1.4	0.6	1.1	1.0	2.4	2.1	2.8	3.7	2.5	2.0	4.9	4.2	5.8	2.9	[illegible]	1.1	[illegible]	[illegible]

Phenological stages

Indian mustard Raya *Brassica juncea* L.	Sowing & seedling establishment (15-21 days)	Vegetative (25-30 days)	Flowering (15-20 days)	Pod development (25-30 days)	Maturity (15-30 days)

FAVOURABLE WEATHER CONDITIONS FOR BETTER YIELD

T_{Max}(°C)	Mean optimum 15-20	15-25	15-20	18-28	20-30
T_{Min}(°C)	Soil temp. > 10	7-10	5-10	8-12	12-15
RH_{M}(%)	>60	>65	>65	>90	-
RH_{E} (%)	>35	>30	>35	>40	-
WS (kmph)	-	-	3-7	3-6	3-6
BSS (hrs)	Clear days	Clear days	>8 hrs	>9 hrs	Clear days
RAINFALL (mm)	-	10-20	20-30	20-25	-

CONGENIAL WEATHER CONDITIONS FOR INCIDENCE OF PESTS & DISEASES

PESTS & DISEASES	Mainly dry weather T_{mean} 27-29 °C RH_{mean} >75 %	Cloudy weather T_{mean} 10-20 °C RH > 82%	T_{max} 18-27 °C, T_{min} 8-12 °C, T_{mean}>10 °C, Monthly RH > 92% (RH_{E} 40% & RH_{mean} 70%)	T_{max}>26 °C, T_{min} 8-16 °C, RH_{mean} 38-87%
Growth phase/period of occurrence	White fly (Seedling establishment)	Aphid [Flowering & pod development (Jan-Feb)]	Alternaria Blight, Downy mildew and White rust (Reproductive & seed development)	Painted bug (Seedling establisment & maturity)

Fig. 15.1: Crop weather calendar for mustard

Information contained in the calendars gives broad indications of the direction of development which may prove useful to the planners, agricultural administrators, plant breeders and the farmers in formulating policy matters regarding plant breeding, crop adoption, drought, supplemental irrigation, maximizing the yield. Crop weather calendars can act as guiding tool while issuing crop weather advisories for the farming community.

15.1. Climate Normals

Climate normals provide a baseline for understanding and managing climate-related risks and opportunities in agriculture, enabling more effective planning and decision-making. The climate normals for major crops grown in India are described below:

15.1.1. Rice

Rice is a tropical (Kharif) crop that thrives in hot and humid climates.

- **Temperature:** Requires high temperatures, ideally above 25°C, throughout its life cycle. The optimum range is often cited as 21–37°C. For germination, the ideal temperature is 20–25°C. During blooming, 26.5–29.5°C is preferred, and for ripening, 20–25°C.

- **Rainfall:** High humidity and an annual rainfall of over 100 cm are essential. It can also be grown in areas with less rainfall with the help of irrigation. Some sources mention an average annual rainfall requirement of around 150 cm.
- **Sunlight:** Prolonged sunshine is beneficial.
- **Soil:** It grows well in a variety of soils like silts, loams, and gravels. Clayey loam soil, which can be easily converted into mud and retain water, is considered ideal. Rice can tolerate both acidic and alkaline soils.
- **Other:** Rice cultivation often requires semi-aquatic conditions, with fields flooded with 10-12 cm of water during sowing.

15.1.2. Wheat

Wheat is a temperate (Rabi) crop that requires a cool climate with moderate rainfall.

- **Temperature:** A moderately cool climate is needed. At the time of sowing, 10-15°C is ideal, while 21-26°C is preferred during ripening and harvesting. The optimum temperature range for germination is 20–25°C.
- **Rainfall:** Requires about 75-100 cm of well-distributed rainfall.
- **Sunlight:** Warm and sunny weather is essential at the time of ripening.
- **Soil:** Well-drained, fertile loamy and clayey loamy soils are best suited. It can also grow in the black soil of the Deccan plateau.
- **Other:** Drizzles and cloudiness during ripening can increase yield. Frost or hailstorms during the flowering period can be harmful.

15.1.3. Sugarcane

Sugarcane is a tropical crop that requires a hot and humid climate.

- **Temperature:** Requires a hot climate, ideally between 21–27°C. Temperatures above 20°C in the second half of the crop season help in juice acquisition and thickening. The optimum temperature for cane growth (germination) is 27–33°C.
- **Rainfall:** Needs 75-150 cm of rainfall. A total rainfall between 1100 and 1500 mm with proper distribution (abundant during vegetative growth followed by a dry period for ripening) is considered adequate.
- **Sunlight:** Sugarcane is a sun-loving plant; 7-9 hours of bright sunshine is very useful for growth and ripening.

- **Soil:** Can tolerate any soil that retains moisture. Deep, rich loamy soils are often mentioned as suitable. Well-drained soils are preferred as waterlogging can be detrimental.
- **Other:** A cool, dry winter during ripening and harvesting is ideal. Frost is harmful.

15.1.4. Pearl Millet (Bajra)

Pearl millet is a hardy crop well-suited to hot and dry conditions.

- **Temperature:** Optimum growth occurs between 27–28°C. It can tolerate temperatures from 15°C to 40°C. The minimum soil temperature for germination is around 18°C (65°F).
- **Rainfall:** Requires low to moderate rainfall, typically between 40-50 cm annually. It is drought-tolerant and cannot withstand high rainfall (90 cm or above).
- **Soil:** Grows best in light, well-drained soils with low inherent fertility and mild salinity. It does not tolerate acidic soils or waterlogging.
- **Other:** It is often grown in rainfed areas. Critical growth stages for irrigation (if available) are maximum tillering, flowering, and grain filling.

15.1.5. Sorghum (Jowar)

Sorghum is also a warm-weather crop known for its drought tolerance.

- **Temperature:** Requires warm temperatures for good germination and growth, with an optimum range of 26–30°C. The minimum temperature for seed germination is 7–10°C, and a soil temperature of 17–18°C is ideal for seedling emergence.
- **Rainfall:** Best suited for areas with an average rainfall between 450-650 mm.6 It can withstand drought and remain dormant under extreme stress, recovering with rains.
- **Soil:** Grows well in a wide range of soils, ideally well-drained fertile soils with a pH between 6.0 and 7.5.
- **Other:** Sorghum is a short-day plant, requiring long night hours before the reproductive stage. Most improved varieties for India are not very sensitive to photoperiod.

15.1.6. Finger Millet (Ragi)

Finger millet is a hardy crop adapted to dry weather and higher elevations.

- **Temperature:** Requires a temperature range of 20–34°C. Sowing temperature is ideally 30–34°C, and harvesting temperature 20–30°C.

Optimum day temperature is around 32°C and night temperature around 25°C.

- **Rainfall:** Suited for areas with an annual rainfall of around 100 cm. It can grow under rainfed as well as irrigated conditions.
- **Soil:** Can be grown on a variety of soils from rich loam to poor shallow upland soils with good organic matter. Well-drained black soils are also suitable. The ideal soil pH is between 4.5 and 8.0.
- **Other:** It can withstand severe drought conditions. In high rainfall areas, it's often grown as a transplanted crop on well-drained soils.

15.1.7. Potatoes (Aloo)

Potatoes are a temperate climate crop that grows under a diverse range of conditions where the growing season is moderately cool.

- **Temperature:** Vegetative growth is best at 24°C, while tuber development is favored at around 20°C. A temperature range of 14–25°C is generally suitable.
- **Rainfall:** Requires about 300-500 mm of rainfall.
- **Soil:** Can grow on a wide range of soils, but well-drained sandy loam and medium loam soils rich in organic content are best. The ideal soil pH is 5.2-6.4. It cannot grow in waterlogged, saline, or alkaline soils.
- **Other:** Grown as a summer crop in the hills and as a winter crop in tropical and subtropical regions.

15.1.8. Groundnut (Moongphali)

Groundnut thrives in tropical, subtropical, and warm temperate climates.

- **Temperature:** Requires warm temperatures throughout its growth. Optimum temperatures for vegetative and reproductive growth are around 30°C.
- **Rainfall:** Primarily grown in rainfed conditions. Annual rainfall in groundnut-growing areas in India varies from 400-1500 mm, usually received within 2-4 rainy months.
- **Soil:** Grown in soils ranging from coastal sands to Vertic Inceptisols in India, and Alfisols and Oxisols in Africa. Well-drained, light-textured, loose, friable sandy loam or sandy clay loam soils are most suitable.
- **Other:** Groundnut is unique as it's cultivated under different production systems/seasons: Kharif (rainy), Rabi (winter/post-rainy), summer, and spring. Timely sowing with the onset of monsoon is crucial.

15.1.9. Sesame (Til)

is a warm-season crop that thrives in tropical and subtropical regions.

- **Temperature:** Requires a high and constant temperature, ideally in the range of 26–30°C for optimum growth. Soil temperature should be above 25°C for germination. Temperatures above 40°C can repress pollination and capsule formation.
- **Rainfall:** Optimal rainfall is between 500-800 mm annually. It does not tolerate excess water and is susceptible to waterlogging.
- **Soil:** Grows best on medium to light, well-drained soils. Prefers a pH between 5-8. Does not tolerate salinity or standing water.
- **Other:** High humidity can damage the plant, exposing leaves and pods to blight and fungal diseases.

15.1.10. Sunflower (Surajmukhi)

Sunflower is a versatile crop grown in a wide range of climates.

- **Temperature:** Mean daily temperatures for good growth are between 18–25°C. It requires a low temperature (3–6°C) during germination and an optimum of 20–26°C later on. Temperatures above 40°C can cause desiccation of pollen.
- **Rainfall:** Water requirements vary from 600 to 1000 mm depending on the climate and growing period.
- **Soil:** Grows well on neutral to moderately alkaline soils with good drainage and some sand content is preferred. The ideal pH range is 6.5 - 8.0.
- **Other:** Warm weather and non-cloudy sunny days are favorable during the reproductive phase (flowering to maturity). It is susceptible to frost.

15.1.11. Soybean

Soybean is grown in warm conditions across tropical, subtropical, and temperate climates.

- **Temperature:** Optimum temperature for most varieties is 26–30°C. Growth rates decrease above 35°C and below 18°C. Soil temperatures of 15.5°C or above favor rapid germination.
- **Rainfall:** Requires 400 to 500 mm of water in a season for a good crop, with high moisture needs during germination, flowering, and pod formation. Dry weather is needed for ripening.
- **Soil:** Performs poorly on sandy soils and soils with low water storage. Medium-textured (silt loam) soils are better. Susceptible to waterlogging between emergence and the four-leaf stage.

- **Other:** Day length is a key factor for most varieties as they are short-day plants.

15.1.12. Cotton (Kapas)

Cotton is a semi-xerophyte grown in tropical and subtropical conditions.

- **Temperature:** A minimum temperature of 15°C is required for germination. Optimum for vegetative growth is 21–27°C, and it can tolerate up to 43°C. Warm days and cool nights during fruiting are beneficial.
- **Rainfall:** Requires 500-650 mm of rainfall. Heavy rainfall during the early stage is undesirable, while dry autumn months are preferred for good quality.
- **Soil:** Grown on a variety of soils, from well-drained deep alluvial soils to black clayey soils. Prefers well-drained soils and is sensitive to waterlogging.
- **Other:** Requires 4-5 months of uniformly high temperatures during its growth period.

15.1.13. Pigeonpea (Arhar/Tur)

Pigeonpea is mainly cultivated in the semi-arid tropics.

- **Temperature:** Can be grown with temperatures ranging from 26–30°C in the rainy season (June to October) and 17–22°C in the post-rainy season (November to March).
- **Rainfall:** Grown in regions with 600-1400 mm of annual rainfall and a growing period of 90-180 days.
- **Soil:** Successfully grown in black cotton soils that are well-drained with a pH ranging from 7.0-8.5.
- **Other:** Sensitive to low radiation at pod development; flowering during monsoon and cloudy weather can lead to poor pod formation.

Climate normals for Pea, Chickpea, Mustard, Green gram, and Cowpea:

15.1.14. Pea (Matar)

Pea is a cool-season crop.

- **Temperature:** Grows best in cool temperatures, ideally between 15–25°C. High temperatures (>30°C) can be detrimental, especially during flowering and pod development. For germination, the optimum soil temperature is around 10–18°C.

- **Rainfall:** Requires moderate rainfall, typically between 60-100 cm, well-distributed throughout the growing season. Excessive rainfall can lead to waterlogging and diseases.
- **Soil:** Prefers well-drained, loamy soils rich in organic matter. The ideal soil pH is between 6.0 and 7.5.
- **Other:** Requires a relatively cool and moist growing season. Often grown as a winter or spring crop.

15.1.15. Chickpea (Chana)

Chickpea is a rabi (winter) season crop that prefers cool and dry conditions.

- **Temperature:** Requires cool temperatures during the vegetative phase and warm temperatures during maturity. The optimum temperature range for growth is 15–20°C. High temperatures at the time of flowering can cause flower drop.
- **Rainfall:** Mostly grown as a rainfed crop, requiring about 40-50 cm of rainfall. Excessive rainfall, especially during flowering and pod development, is harmful.
- **Soil:** Well-drained, light to medium textured soils are preferred. It can also grow on black cotton soils with good drainage. The ideal soil pH is 5.5 to 7.5.
- **Other:** Clear sunny weather is beneficial for its growth.

15.1.16. Mustard (Sarson)

Mustard is a rabi (winter) season crop that thrives in cool climates.

- **Temperature:** Requires a cool climate during the growing period. Optimum temperature for germination is 15–20°C, and for growth is 20–25°C. High temperatures at the time of flowering can reduce yield.
- **Rainfall:** Requires about 30-40 cm of rainfall, mainly during the vegetative growth phase. Rainfall at the time of flowering and pod formation can be detrimental.
- **Soil:** Grows well on a variety of soils, including loamy, sandy loam, and clay loam soils, provided they are well-drained. The ideal soil pH is between 6.0 and 7.5.
- **Other:** Clear sunny weather is conducive for good yield.

15.1.17. Green gram (Moong)

Green gram is a warm-season legume.

- **Temperature:** Requires warm and humid climate. The optimum temperature range for growth is 30–35°C. It is sensitive to low temperatures.
- **Rainfall:** Requires moderate rainfall, around 60-75 cm, well-distributed throughout its short growing season. It is susceptible to waterlogging.
- **Soil:** Grows well on well-drained loamy to sandy loam soils. The ideal soil pH is between 6.0 and 7.5.
- **Other:** It's often grown as a short-duration crop in different seasons.

15.1.18. Cowpea (Lobia)

Cowpea is a warm-season legume.

- **Temperature:** Requires warm temperatures for growth, with an optimum range of 27–30°C. It is sensitive to frost.
- **Rainfall:** Adaptable to a wide range of rainfall conditions, from 40 cm to 120 cm. However, it grows best with 50-75 cm of well-distributed rainfall.
- **Soil:** Can grow on a variety of soils but prefers well-drained sandy loam to loamy soils. Tolerates slightly acidic to slightly alkaline soils (pH 5.5-7.0).
- **Other:** It is relatively drought-tolerant.

15.2. Climate Normals of Livestock in India

It's important to understand that "climate normals" for livestock are less about specific temperature and rainfall ranges like crops, and more about the **thermal comfort zone** where the animals thrive. Outside this zone, they experience stress, which can impact their health and productivity. This comfort zone can vary based on breed, age, physiological state (e.g., lactation), and acclimatization.

15.2.1. Cows and Buffaloes

- **Thermal Comfort Zone:** Generally, cattle and buffaloes in temperate regions have a comfort zone between 10°C and 25°C. However, in India's tropical and subtropical climate, indigenous breeds and buffaloes have adapted to higher temperatures. The **Temperature Humidity Index (THI)** is often used, where a THI above 72 indicates mild stress, 72-78 moderate stress, and above 78 severe stress. Many parts of India experience THI above 72 for a significant part of the year.

- **Impact of Climate:** High temperatures and humidity can lead to heat stress, reducing feed intake, milk production, and reproductive efficiency. Indigenous breeds are generally more heat-tolerant than exotic breeds. Buffaloes, especially swamp buffaloes, are often considered more tolerant to high temperatures than European cattle, although they can still experience heat stress. They often rely on access to water for cooling.

15.2.2. Goats and Sheep

- **Thermal Comfort Zone:** Goats and sheep are generally more adaptable to a wider range of temperatures than cattle. Their comfort zone can range from 7°C to 27°C, but they can tolerate colder and warmer temperatures depending on breed and wool/hair coat. In hot arid and semi-arid regions of western India, they are particularly vulnerable to high temperatures. A THI above 80 is considered moderate heat stress for sheep and goats.
- **Impact of Climate:** Extreme heat can reduce body weight, growth rates, and reproductive performance in sheep and goats.They employ physiological mechanisms like sweating and increased respiration to cope. Indigenous breeds often have better heat tolerance. Goats, in particular, exhibit resilience to climatic variations due to physiological and behavioral traits like seeking shade.

15.2.3. Pigs

- **Thermal Comfort Zone:** Pigs are quite sensitive to temperature extremes because they lack functional sweat glands. Their ideal temperature range is between 15°C to 21°C. Temperatures above this can lead to heat stress quickly.
- **Impact of Climate:** High temperatures cause panting, reduced feed intake, and decreased growth rates. In hot climates, cooling systems like evaporative pads and misters are crucial. In colder climates, they require protection from the cold to avoid increased feed consumption for maintaining body temperature.

15.2.4. Poultry Birds

- **Thermal Comfort Zone:** The thermoneutral zone for poultry varies with age. For broilers, it starts high (32-34°C on day 1) and gradually decreases to around 20°C by week 4. For layers, the most efficient temperatures are between 20-24°C.
- **Impact of Climate:** Poultry are highly susceptible to heat stress at high temperatures (>24°C), which can reduce feed intake, growth rate, egg production, and egg quality. High humidity exacerbates heat stress. In

cold climates, they need adequate housing and sometimes supplemental heat to maintain body temperature and productivity.

It's important to note that these are general guidelines. The specific requirements and tolerance levels can vary greatly depending on the breed and local environmental conditions in different parts of India.

15.3. Temperature Humidity Index for Livestock

The **Temperature Humidity Index (THI)** is a single value that combines the effects of air temperature and relative humidity to indicate the level of heat stress an animal is likely to experience. It's a useful and relatively easy way to assess the risk of heat stress in livestock.

15.3.1. How is THI Calculated?

There are several formulas for calculating THI. One of the most common for livestock is:

$THI=(1.8\times Tdb+32)-[(0.55-0.0055\times RH)\times(1.8\times Tdb-26)]$

Where:

- Tdb = dry-bulb temperature in degrees Celsius (°C)
- RH = relative humidity expressed as a percentage (e.g., 70 for 70%)

Another simpler formula sometimes used is:

$THI=0.8\times T+(RH/100)\times(T-14.4)+46.4$

Where:

- T = ambient or dry-bulb temperature in °C
- RH = relative humidity as a percentage

15.3.2. THI Requirements/Thresholds for Livestock

Keep in mind that these are general guidelines, and specific breeds, physiological states, and acclimatization can influence an animal's tolerance.

- **Cattle**
  - **No heat stress:** THI < 68
  - **Mild heat stress:** 68 - 71
  - **Mild to moderate heat stress:** 72 - 796
  - **Moderate to severe heat stress:** 80 - 90
  - **Severe heat stress:** >90

Modern, high-producing dairy cows may experience mild heat stress starting at a lower THI (around 68 or even lower according to some research).

- **Goats and Sheep:** THI thresholds for goats and sheep are less well-defined compared to cattle. However, a general guideline suggests:
  - **No heat stress:** THI < 75
  - **Mild heat stress:** 75 - 80
  - **Moderate heat stress:** 80 - 85
  - **Severe heat stress:** >85
- **Pigs:** Pigs are quite sensitive to heat stress. General THI thresholds are:
  - **Comfort:** THI < 70
  - **Mild stress:** 70 - 75
  - **Moderate stress:** 75 - 80
  - **Severe stress:** >80
- **Poultry:** THI thresholds for poultry also vary with age and type (broilers vs. layers):
  - **Comfort:** THI < 72
  - **Light discomfort:** 73 - 76
  - **Moderate discomfort:** 77 - 80
  - **Severe discomfort:** 81 - 84
  - **Life-threatening:** >85

It's crucial to monitor THI, especially during hot and humid periods, and implement strategies to mitigate heat stress in livestock when THI exceeds these thresholds. These strategies can include providing shade, increasing ventilation, providing access to water, and using cooling systems.

16

Weather Forecasting

From the farmer deciding when to sow seeds to the traveler planning their journey, from the pilot navigating the skies to the energy provider managing demand, the state of the atmosphere profoundly influences our daily lives and myriad human activities. The ability to anticipate future atmospheric conditions, a practice we know as **weather forecasting**, is therefore not merely an academic pursuit but a practical necessity (Fig. 16.1).

Fig. 16.1: Weather forecasting

This chapter will delve into the fascinating world of weather forecasting, exploring its historical roots, the fundamental scientific principles that underpin it, and the sophisticated tools and techniques employed by modern meteorologists. We will journey through the process of observing the current state of the atmosphere, the computational power used to model its complex dynamics, and the various ways forecasts are disseminated and interpreted.

Understanding the science behind weather forecasting empowers us to appreciate its capabilities and limitations. It allows us to make more informed decisions in the face of an ever-changing atmospheric environment. Join us as we unravel the methods used to predict the weather, from simple observations to cutting-edge numerical models, and gain insight into the continuous advancements shaping the future of this vital scientific discipline.

16.1. Types of Weather Forecast

There are several ways to categorize weather forecasts, most commonly by the **time range** they cover. Here are the main types:

16.1.1. Nowcasting

- **Time Range:** Present to 0-6 hours ahead. Some definitions narrow this to within the next 2 hours.
- **Focus:** Detailed description of the current weather and very short-term predictions of its evolution.
- **Methods:** Relies heavily on real-time observations from radar, satellites, and surface stations, extrapolated forward in time. Often incorporates mesoscale numerical weather models.
- **Usefulness:** Critical for issuing immediate warnings for severe weather (e.g., thunderstorms, heavy rainfall), aviation, and short-term planning of outdoor activities.
- **Example:** Predicting that a current heavy rain shower will continue to move northeast over the next hour, affecting a specific locality.

16.1.2. Short-Range Forecast

- **Time Range:** 1 to 3 days (up to 72 hours).
- **Focus:** Predicting day-to-day weather conditions, including temperature, precipitation, wind, and cloud cover.
- **Methods:** Primarily based on numerical weather prediction (NWP) models, which use computer simulations of the atmosphere. Meteorologists analyze and interpret model outputs, incorporating their expertise and local knowledge.
- **Usefulness:** Essential for daily planning, agriculture, transportation, and many other sectors.
- **Example:** Forecasting sunny skies and a high of 30°C for tomorrow.

16.1.3. Medium-Range Forecast

- **Time Range:** 3 to 10 days. Some sources extend this up to 2 weeks.
- **Focus:** Providing a general overview of weather trends beyond the immediate short-term. Accuracy decreases with increasing time.
- **Methods:** Primarily relies on global NWP models. Ensemble forecasting (running multiple simulations with slightly different initial conditions) becomes more important to assess the range of possible outcomes and uncertainty.

- **Usefulness:** Helpful for medium-term planning in various industries, including agriculture, energy, and logistics.
- **Example:** Indicating a likelihood of a wet spell arriving in the region around the end of next week.

16.1.4. Extended-Range Forecast

- **Time Range:** Beyond 10 days and up to a month (some define it as between medium-range and long-range, roughly days 10 to 30).
- **Focus:** Identifying broad-scale weather patterns and anomalies (e.g., above or below average temperatures, wetter or drier than normal conditions).
- **Methods:** Relies on global climate models and statistical methods that consider large-scale patterns like sea surface temperatures and atmospheric oscillations.
- **Usefulness:** Useful for long-term planning and for anticipating potential deviations from seasonal norms.

16.1.5. Long-Range Forecast (Seasonal Outlook)

- **Time Range:** One month to several seasons (e.g., 3-6 months or even a year).
- **Focus:** Predicting overall trends and probabilities of certain conditions (e.g., a warmer than average winter, a wetter than average monsoon). Does not provide day-to-day specifics.
- **Methods:** Based on complex global climate models that simulate the interactions between the atmosphere, oceans, land, and ice. Statistical analysis of historical data and teleconnections (long-distance relationships between weather patterns) are also used.
- **Usefulness:** Crucial for strategic planning in agriculture, water resource management, and energy sectors.
- **Example:** Forecasting a higher probability of a warmer and drier than normal summer for a particular region.

Additionally, forecasts can be categorized by their **purpose** or the **phenomena** they predict, such as:

- **Severe Weather Forecasts/Warnings:** Focused on hazardous weather like hurricanes, tornadoes, blizzards, and heatwaves.
- **Aviation Forecasts:** Tailored for the needs of the aviation industry, including wind, visibility, and icing conditions.

- **Marine Forecasts:** Providing information relevant to marine activities, such as wave height, sea state, and winds.
- **Agricultural Forecasts (Agromet Advisories):** Specific to farming, including rainfall, temperature, humidity, and their impact on crops.

The time-range categorization is the most common way to differentiate types of weather forecasts.

16.2. Weather Forecasting for Agriculture

Weather plays a pivotal role in agriculture, influencing everything from planting and growth to harvesting and the prevalence of pests and diseases (Fig. 16.2).

Fig. 16.2: Weather forecasting for agriculture

Accurate and timely weather forecasts tailored to agriculture are therefore indispensable for farmers to make informed decisions, optimize productivity, and mitigate risks. This specialized area of weather forecasting is often referred to as **agrometeorological forecasting** or the provision of **agromet advisories**.

16.2.1. Key Features of Agricultural Weather Forecasts

Agricultural weather forecasts go beyond general public forecasts by often including parameters critical for farming operations:

- **Rainfall:** Amount, intensity, timing, and spatial distribution are crucial for irrigation scheduling, planting, harvesting, and assessing the risk of floods or droughts.
- **Temperature:** Minimum and maximum temperatures affect crop development stages, the likelihood of frost damage, and the activity of pests and diseases. Soil temperature is also important for germination.
- **Humidity:** High humidity can favor the spread of fungal diseases, while low humidity can increase evapotranspiration rates and the need for irrigation.

- **Wind:** Speed and direction are important for planning pesticide and fertilizer applications (to avoid drift), assessing evapotranspiration, and understanding the potential for lodging (falling over) in crops.
- **Solar Radiation:** Affects photosynthesis and crop growth rates.
- **Evapotranspiration (ET):** An estimate of water loss from the soil and plants, vital for irrigation management.
- **Dew Point:** Indicates the likelihood of dew formation, which can influence the onset and spread of certain plant diseases.
- **Extreme Weather Events:** Warnings for frost, heat waves, hailstorms, strong winds, and other severe weather that can damage crops or livestock.

16.2.2. Types of Agricultural Weather Forecasts (by Time Range)

Similar to general weather forecasts, agricultural forecasts are also categorized by their lead time:

- **Nowcasting (0-6 hours):** Crucial for immediate decisions like whether to proceed with spraying pesticides or to take shelter from an approaching thunderstorm.
- Relies heavily on real-time data.
- **Short-Range (1-3 days):** Helps in planning daily farm activities such as irrigation, fertilizer application, and scheduling labor.
- **Medium-Range (3-10 days):** Useful for more strategic planning, like deciding on the timing of sowing or harvesting, and managing water resources over the coming week.
- **Extended-Range (10 days to a month):** Provides insights into broader trends, assisting in decisions about crop management and resource allocation over a longer period.
- **Long-Range/Seasonal Outlook (1 month to several seasons):** Invaluable for making decisions about which crops to plant in a particular season, planning for potential water availability (especially important for monsoon-dependent regions like India), and anticipating potential pest or disease outbreaks based on expected seasonal conditions.

16.2.3. Methods and Tools for Agricultural Weather Forecasting

Generating these specialized forecasts involves a combination of:

- **Traditional Meteorological Observations:** Data from surface weather stations, radiosondes (weather balloons), buoys, and aircraft provide the foundational information.

- **Satellite Imagery:** Offers a broad view of cloud cover, vegetation health, and land surface temperature, which are all relevant to agriculture.
- **Radar:** Detects precipitation intensity and movement, crucial for short-range warnings of heavy rainfall or hail.
- **Numerical Weather Prediction (NWP) Models:** Sophisticated computer models that simulate the atmosphere based on physical laws.
- These models are increasingly incorporating land surface processes relevant to agriculture, such as soil moisture.
- **Agrometeorological Models:** These models go a step further by integrating weather data with information about specific crops (their growth stages, water requirements, susceptibility to pests, etc.) to provide tailored advisories.
- **Statistical Methods:** Analyzing historical weather data and crop yields to identify relationships and make probabilistic forecasts.
- **Artificial Intelligence (AI) and Machine Learning (ML):** Increasingly used to analyze large datasets, improve the accuracy of forecasts, and develop more localized predictions.

16.2.4. Agrometeorological Advisory Services

In many countries, including India, weather forecasts are translated into actionable advice for farmers through **Agrometeorological Advisory Services (AAS)**. These services typically involve:

- **Preparing Agromet Bulletins:** These bulletins, often issued 2-3 times a week, contain weather forecasts relevant to agriculture for a specific region, along with crop-specific advisories.
- **Crop-Specific Recommendations:** Based on the forecast, advisories might suggest the optimal timing for sowing, irrigation, application of fertilizers and pesticides, harvesting, and other farm operations.
- **Pest and Disease Warnings:** Predicting conditions favorable for the outbreak of pests and diseases and advising on preventive or control measures.
- **Warnings for Extreme Weather:** Alerting farmers to impending severe weather events and suggesting measures to protect their crops and livestock.
- **Dissemination:** Advisories are often disseminated through various channels like mobile SMS, radio, television, newspapers, and increasingly through dedicated mobile apps and web portals in local languages. In India, the India Meteorological Department (IMD) and agricultural universities play a significant role in this.

16.2.5. Importance of AAS in Agriculture

Accurate agricultural weather forecasts and effective agromet advisory services can lead to:

- **Optimized Planting and Harvesting:** Aligning these activities with favorable weather conditions.
- **Efficient Water Management:** Scheduling irrigation based on predicted rainfall and evapotranspiration.
- **Timely Application of Inputs:** Applying fertilizers and pesticides when they will be most effective and least likely to be washed away.
- **Reduced Crop Losses:** Taking protective measures against adverse weather, pests, and diseases.
- **Improved Resource Use Efficiency:** Making better decisions regarding labor, machinery, and other resources.
- **Enhanced Productivity and Income:** Ultimately leading to higher yields and better economic returns for farmers.

In conclusion, weather forecasting for agriculture is a specialized and vital service that combines meteorological science with agricultural knowledge to support farmers in making weather-smart decisions throughout the crop cycle. As technology advances, these forecasts are becoming more accurate, localized, and tailored to the specific needs of the agricultural community.

16.3. Weather Warnings and Alerts for Agriculture

Weather warnings and alerts are crucial for farmers as they help protect crops, livestock, and farm infrastructure from potentially damaging weather events (Fig. 16.3). These warnings enable farmers to take proactive measures to minimize losses and ensure their safety.

Fig. 16.3: Weather warnings and alerts for agriculture

Some common types of weather warnings & alerts relevant to agriculture

16.3.1. Rainfall Warnings

- **Heavy Rainfall Warning:** Alerts farmers to the possibility of significant rainfall that could lead to waterlogging in fields, soil erosion, and damage to crops, especially during flowering or harvesting stages. Advisories might include draining excess water, postponing irrigation, or harvesting mature crops quickly.
- **Excessive Rainfall/Flash Flood Warning:** Indicates a high risk of sudden and intense rainfall that can cause widespread flooding, damaging crops, washing away topsoil, and posing a threat to livestock and farm buildings. Farmers might be advised to move livestock to higher ground, secure farm equipment, and avoid low-lying areas.
- **Deficient Rainfall/Drought Alert:** Warns of prolonged dry spells that could lead to water stress in crops, reduced yields, and scarcity of water for irrigation and livestock. Advisories could include implementing water-saving irrigation techniques, selecting drought-tolerant crops, or managing livestock water resources carefully.

16.3.2. Temperature Warnings

- **Frost Warning:** Alerts farmers to the risk of freezing temperatures that can damage sensitive crops, especially during early growth stages or flowering. Protective measures like covering plants, using smudge pots, or irrigating before frost might be recommended.
- **Heat Wave Warning:** Indicates dangerously high temperatures that can cause heat stress in crops and livestock, leading to reduced productivity and even mortality. Farmers might be advised to provide shade and ample water for livestock, adjust irrigation schedules for crops, and avoid working outdoors during peak heat hours.

16.3.3. Wind Warnings

- **Strong Wind Warning:** Alerts to the possibility of high winds that can cause lodging in crops (especially tall ones like maize or sugarcane), damage to young plants, and make spraying of pesticides or fertilizers ineffective. Farmers might be advised to provide support to vulnerable crops or postpone spraying.
- **Storm/Cyclone Warning:** Indicates the potential for very strong winds and heavy rainfall associated with storms or cyclones, which can cause widespread damage to crops, trees, and farm infrastructure. Farmers would typically be advised to secure buildings, move livestock to safety, and take shelter.

16.3.4. Hail Warnings

- **Hailstorm Warning:** Alerts to the risk of hailstorms that can cause significant physical damage to crops, fruits, and vegetables. Protective measures are limited but might include using netting in some high-value crop situations.

16.3.5. Other Warnings

- **Thunderstorm and Lightning Warnings:** Primarily for the safety of farmers and farmworkers, advising them to seek shelter to avoid lightning strikes.
- **Pest and Disease Outbreak Warnings:** While not strictly weather warnings, these are often linked to weather conditions favorable for pest and disease proliferation. Farmers are alerted to monitor their crops and take preventative measures.

16.3.6. Dissemination of Warnings and Alerts

Timely dissemination is crucial for these warnings to be effective. Common methods include:

- **Mobile SMS alerts:** Direct messages to farmers' mobile phones.
- **Radio and Television broadcasts:** Reaching a wider audience, especially in rural areas.
- **Dedicated mobile apps:** Providing detailed weather information and advisories. Examples in India include 'Meghdoot' and 'Mausam'.
- **Web portals:** Offering comprehensive weather information and agricultural advisories. The India Meteorological Department (IMD) provides these services through its website & the 'Mausamgram' portal.
- **Kisan Call Centres:** Providing information and answering queries in local languages.
- **Social media platforms:** Used by some Agromet Field Units (AMFUs) to disseminate information quickly.
- **Gram Panchayat Level Weather Forecasting (GPLWF):** Initiatives to provide forecasts at the village level through platforms like e-Gramswaraj and the Meri Panchayat app.

In India, the **India Meteorological Department (IMD)** plays a central role in generating weather forecasts and warnings, which are then used by Agromet Field Units (AMFUs) to prepare and disseminate crop-specific advisories. These advisories often include warnings about impending adverse weather and suggest actions farmers should take.

17

Climate Change and Agriculture

Agriculture, the cultivation of plants and rearing of animals for food, fiber, and other products, is intrinsically linked to the Earth's climate. Temperature, rainfall patterns, and atmospheric composition profoundly influence agricultural productivity and distribution. However, the stability of these climatic conditions is increasingly threatened by **climate change**, driven primarily by the escalating concentration of greenhouse gases in the atmosphere. This chapter delves into the multifaceted relationship between climate change and agriculture, exploring the ways in which altered climatic conditions affect agricultural practices, productivity, and sustainability. We will examine the impact of rising temperatures, shifting precipitation patterns, and increased frequency of extreme weather events on crop yields, livestock health, and overall agricultural output. Furthermore, the chapter will discuss the implications of these changes for global food security, economic stability, and the livelihoods of millions of people who depend on agriculture. Finally, we will explore strategies for mitigating the adverse effects of climate change on agriculture and enhancing the resilience of agricultural systems in the face of this unprecedented challenge.

17.1. Climate Variability and Climate Change

It's important to understand the distinction between climate variability and climate change, as they describe different aspects of how the Earth's climate system behaves.

17.1.1. Climate Variability

Climate variability refers to natural fluctuations in climate conditions around a long-term average. These fluctuations can occur on various timescales, including:

- **Seasonal:** Changes in temperature and precipitation between summer and winter.
- **Year-to-year:** Variations like El Niño and La Niña events, which can cause shifts in weather patterns across the globe.
- **Decadal:** Longer-term patterns, such as changes in ocean currents.

These variations are caused by natural processes within the Earth's climate system, such as:

- **Ocean-atmosphere interactions:** The exchange of heat and moisture between the oceans and the atmosphere.
- **Solar activity:** Changes in the amount of energy emitted by the sun.
- **Volcanic eruptions:** Large eruptions can release particles into the atmosphere, affecting temperatures.

17.1.2. Climate Change

Climate change, on the other hand, refers to a long-term shift in the average climate conditions. This change persists over decades or even longer. The key difference is the directionality and persistence of the change. While climate variability involves fluctuations around an average, climate change involves a fundamental alteration of that average (Fig. 17.1).

Fig. 17.1: Climate change

The primary driver of current climate change is the increase in greenhouse gas concentrations in the Earth's atmosphere, largely due to human activities such as:

- **Burning fossil fuels:** For energy production, transportation, & industry.
- **Deforestation:** Clearing forests reduces the amount of carbon dioxide absorbed by plants.
- **Agriculture:** Certain agricultural practices release greenhouse gases like methane and nitrous oxide.

17.1.3. Key Differences Summarized

Table: Key differences between climate variability and climate change:

Feature	Climate Variability	Climate Change
Timescale	Short-term (seasonal, year-to-year, decadal)	Long-term (decades, centuries, or longer)
Nature of Change	Fluctuations around a long-term average	Persistent shift in the average climate conditions
Primary Causes	Natural processes within the Earth's climate system (e.g., El Niño, volcanic eruptions)	Increase in greenhouse gas concentrations in the atmosphere, primarily due to human activities
Examples	El Niño/La Niña cycles, variations in monsoon patterns, changes in average temperature from one year to the next	Global warming, rising sea levels, changes in global precipitation patterns
Reversibility	Generally more reversible	Less reversible on human timescales

17.2. Global Warming

Global warming refers to the long-term increase in Earth's average surface temperature. It's a key aspect of climate change, and while the planet's temperature has fluctuated naturally in the past, the current warming trend is occurring at an unprecedented rate (Fig. 17.2). Scientists are virtually certain that this is primarily driven by human/anthropogenic activities.

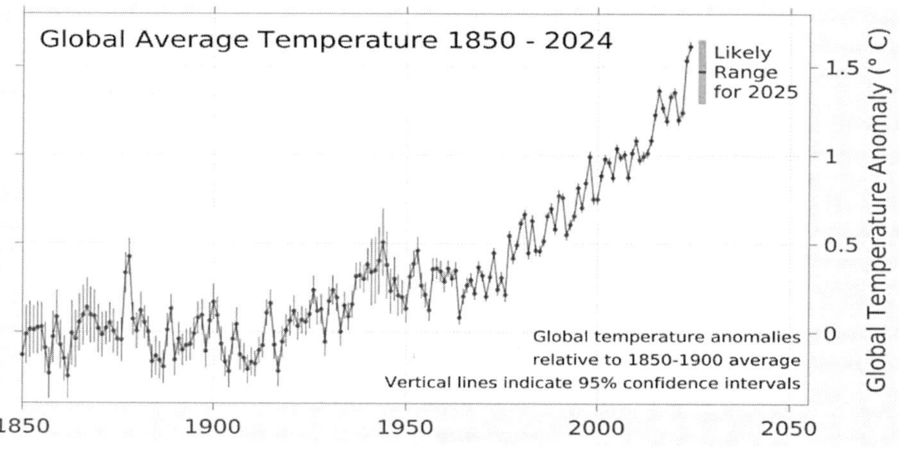

Fig. 17.2: Global warming / average temperature anomalies

17.2.1. The Greenhouse Effect and Global Warming

To understand global warming, it's essential to grasp the **greenhouse effect**. Here's how it works:

- **Solar Radiation:** The Earth receives energy from the sun in the form of solar radiation.
- **Absorption and Reflection:** Some of this radiation is absorbed by the Earth's surface, while some is reflected back into space.
- **Greenhouse Gases:** Certain gases in the Earth's atmosphere, known as greenhouse gases, trap some of the outgoing energy, warming the planet. This is a natural process that keeps the Earth habitable.
- **Human Enhancement:** Human activities are increasing the concentration of these greenhouse gases, trapping more heat and causing the Earth's temperature to rise.

17.2.2. Causes of Global Warming

The primary driver of the current global warming trend is the increase in greenhouse gas concentrations in the atmosphere. The main culprits include:

- **Carbon Dioxide (CO_2):** The most significant greenhouse gas, primarily released through the burning of fossil fuels (coal, oil, and natural gas) for energy production, transportation, and industrial processes. Deforestation also reduces the amount of CO_2 absorbed by trees. The recent daily CO_2 concentration (as on 31st May, 2025) observed was approx. 431 ppm. These daily CO_2 readings are updated automatically with the latest atmospheric CO_2 readings available from the Mauna Loa Observatory in Hawaii (https://www.co2.earth/daily-co2).
- **Methane (CH_4):** A more potent greenhouse gas than CO_2, though it has a shorter lifespan in the atmosphere. Major sources include agricultural activities (especially livestock farming and rice cultivation), natural gas leaks, and the decomposition of organic waste in landfills.
- **Nitrous Oxide (N_2O):** Emitted from agricultural and industrial activities, the combustion of fossil fuels, and the treatment of wastewater.
- **Fluorinated Gases (HFCs, PFCs, SF_6, NF_3):** Synthetic gases used in refrigerants, aerosols, and industrial processes. These are very potent greenhouse gases, with a high global warming potential.

17.2.3. Effects of Global Warming

The consequences of global warming are far-reaching and affect various aspects of the Earth's systems. Here are some of the key effects:

- **Rising Temperatures:** The most direct effect is the increase in average global temperatures. This leads to more frequent and intense heatwaves, affecting human health, agriculture, and ecosystems.

- **Changes in Precipitation Patterns:** Global warming is altering rainfall and snowfall patterns, leading to increased drought in some areas and more intense rainfall and flooding in others.
- **Melting Ice and Rising Sea Levels:** Higher temperatures are causing glaciers and ice sheets to melt at an accelerated rate, and ocean water expands as it warms. This leads to rising sea levels, threatening coastal communities with increased flooding, erosion, and saltwater intrusion into freshwater sources.
- **Ocean Acidification:** The absorption of excess CO_2 by the oceans is causing them to become more acidic, harming marine life, particularly shellfish and coral reefs.
- **Extreme Weather Events:** Global warming is contributing to the increased frequency and intensity of extreme weather events such as hurricanes, cyclones, wildfires, and heatwaves. These events can cause widespread damage to infrastructure, economies, and human lives.
- **Ecosystem Disruption:** Many species are struggling to adapt to the rapidly changing climate, leading to shifts in their geographic ranges, altered migration patterns, and increased risk of extinction.
- **Impacts on Agriculture and Food Security:** Changes in temperature, precipitation patterns, and the increased frequency of extreme weather events can disrupt agricultural production, leading to decreased crop yields, livestock losses, and threats to global food security.
- **Human Health Impacts:** Global warming poses significant risks to human health, including heat-related illnesses, respiratory problems due to increased air pollution, the spread of infectious diseases, and mental health challenges due to displacement and trauma caused by climate-related disasters.

17.2.4. Climate Change impact on Agriculture

Climate change significantly impacts agriculture in numerous ways, threatening food security and the livelihoods of millions of people who depend on farming. Here's a breakdown of the key impacts:

- **Changes in Temperature**
    - **Increased Heat Stress:** Higher temperatures can reduce crop yields, especially if they occur during critical growth stages like flowering and pollination. Livestock are also affected by heat stress, leading to decreased productivity, fertility, and milk production.

- **Lengthened Growing Seasons:** In some regions, warmer temperatures may lengthen the growing season, potentially allowing for new crops or increased yields. However, this can also disrupt traditional farming practices and increase the risk of pest and disease outbreaks.
- **Altered Crop Development:** Temperature changes can affect the timing of crop development, leading to mismatches between crop growth stages and the availability of water and other resources.

- **Changes in Precipitation Patterns**
  - **Drought:** In many areas, climate change is expected to increase the frequency and severity of drought, reducing water availability for irrigation and rainfed agriculture. This can lead to crop failure, livestock losses, and increased competition for water resources.
  - **Increased Flooding:** Conversely, other regions may experience more intense rainfall and flooding, which can damage crops, erode soil, and disrupt planting and harvesting activities.
  - **Erratic Rainfall:** Changes in the timing and distribution of rainfall can make it difficult for farmers to plan their planting and harvesting schedules, leading to decreased yields and increased risk of crop failure.
- **Increased CO_2 Levels**
  - **Increased Photosynthesis:** Higher CO_2 concentrations can increase the rate of photosynthesis in some crops, potentially leading to higher yields. However, this effect may be offset by the negative impacts of higher temperatures and altered precipitation patterns.
  - **Reduced Nutritional Quality:** In some crops, increased CO_2 can reduce the concentration of essential nutrients like protein, zinc, and iron, potentially affecting human health.
- **Sea Level Rise**
  - **Loss of Agricultural Land:** Sea level rise can inundate low-lying coastal areas, leading to the loss of valuable agricultural land.
  - **Saltwater Intrusion:** Saltwater intrusion into freshwater aquifers can make irrigation water too saline for many crops, further reducing agricultural productivity.
- **Increased Frequency and Intensity of Extreme Weather Events**
  - **Heatwaves, Droughts, and Floods:** These events can cause widespread damage to crops and livestock, leading to significant yield losses and economic hardship for farmers.

  - **Stronger Storms:** More intense storms can destroy crops, damage infrastructure, and increase the risk of soil erosion and flooding.
- **Pests and Diseases Dynamics**
  - **Increased Pest and Disease Pressure:** Warmer temperatures and altered precipitation patterns can create more favorable conditions for the survival and spread of pests and diseases, potentially leading to significant crop losses.

17.3. Strategies to Cope with Climate Change in Agriculture

To ensure food security and support the livelihoods of farmers in the face of climate change, it is crucial to adopt a range of adaptation and mitigation strategies. Here are some key approaches:

17.3.1. Adaptation Strategies

Adaptation strategies aim to reduce the vulnerability of agricultural systems to the adverse effects of climate change.

- **Climate-Smart Agriculture (CSA):** CSA practices enhance agricultural productivity, resilience, and adaptive capacity while reducing greenhouse gas emissions. These practices include:
  - **Conservation Agriculture:** Minimizing soil disturbance through no-till farming, cover cropping, and crop rotations to improve soil health, conserve water, and reduce erosion.
  - **Water Management:** Implementing efficient irrigation techniques (e.g., drip irrigation, precision irrigation), rainwater harvesting, and improved drainage systems to conserve water and reduce the risk of drought and flooding.
  - **Diversification:** Diversifying crops, livestock, and farming systems to reduce vulnerability to climate-related risks. This can include planting a mix of crop varieties with different climate tolerances, integrating livestock and crop production, and adopting agroforestry practices.
  - **Improved Crop & Livestock Varieties:** Developing & disseminating crop varieties and livestock breeds that are more tolerant to heat, drought, flooding, pests, and diseases. This can be achieved through conventional breeding, biotechnology, and genetic engineering.
  - **Climate Information Services:** Providing farmers with timely and accurate weather forecasts, early warning systems, and climate projections to help them make informed decisions about planting, harvesting, and other farm operations.

- **Sustainable Land Management:** Practices like terracing, contour farming, and agroforestry help to prevent soil erosion, conserve water, and maintain soil fertility.
- **Disaster Risk Management:** Developing strategies to help farmers prepare for and cope with extreme weather events, such as drought-resistant farming practices, flood-proof infrastructure, and insurance schemes.
- **Policy and Institutional Support:** Governments play a crucial role in supporting adaptation efforts through policies that promote sustainable agriculture, invest in research and development, and provide financial assistance to farmers.

17.3.2. Mitigation Strategies

Mitigation strategies aim to reduce greenhouse gas emissions from agriculture, helping to slow the pace of climate change.

- **Reducing Emissions from Livestock:** Improving livestock management practices to reduce methane emissions from enteric fermentation (e.g., through improved feed quality) and manure management (e.g., through anaerobic digestion).
- **Improving Nitrogen Management:** Optimizing the use of nitrogen fertilizers to reduce nitrous oxide emissions, a potent greenhouse gas. This can involve using precision agriculture techniques, applying fertilizers at the right time and rate, and adopting alternative fertilization methods like legume intercropping.
- **Carbon Sequestration:** Enhancing the capacity of agricultural lands to store carbon in the soil and biomass. Practices that promote carbon sequestration include:
    - **Conservation Agriculture:** No-till farming, cover cropping, and crop rotations increase soil organic carbon.
    - **Agroforestry:** Integrating trees into farming systems to store carbon in tree biomass and soil.
    - **Biochar Application:** Applying biochar (a charcoal-like substance) to the soil can improve soil fertility and sequester carbon.
- **Renewable Energy:** Using renewable energy sources like solar, wind, and biogas to power farm operations, reducing reliance on fossil fuels.

By implementing a combination of adaptation and mitigation strategies, the agricultural sector can become more resilient to the impacts of climate change and contribute to global efforts to reduce greenhouse gas emissions.

18

Weather and Agriculture Relationship

Agriculture, the cultivation of life that sustains human civilization, is profoundly intertwined with the rhythms of the natural world. Of all the environmental factors that influence farming, **weather** reigns supreme, orchestrating a complex symphony of elements that dictate the success or failure of crops. This chapter delves into the intricate relationship between weather and agriculture, exploring how temperature, precipitation, sunlight, and other atmospheric conditions shape every facet of agricultural production.

Weather's influence begins with the very first stage of a plant's life. Temperature governs seed germination and seedling development, while sunlight provides the energy for photosynthesis, the process by which plants convert light energy into chemical energy. The availability of water, dictated by rainfall patterns, is crucial for plant growth and nutrient absorption. Wind patterns can aid in pollination but also cause damage to crops.

Throughout the growing season, weather continues to play a critical role. Optimal temperatures are essential for each stage of plant development, from flowering to fruit production. Adequate rainfall, or irrigation to supplement it, ensures healthy growth and abundant yields.

Extreme weather events, such as droughts, floods, heatwaves, and frosts, can have devastating consequences for agriculture. These events can destroy crops, reduce yields, and disrupt food production, leading to economic losses and food insecurity (Fig. 18.1).

Fig. 18.1: Weather and agriculture relationship

The impact of weather on agriculture extends beyond crop production. It also affects livestock farming, influencing the availability of pasture and fodder, and the health and productivity of animals. Additionally, weather patterns can influence the prevalence of pests and diseases, which can further impact agricultural yields.

Understanding this vital connection is crucial for optimizing agricultural practices, mitigating risks, and ensuring a sustainable future for farming in an ever-changing world. This chapter will explore these aspects in detail, highlighting the importance of weather forecasting, climate change adaptation, and sustainable farming practices in ensuring food security for a growing global population.

18.1. Weather and Agricultural Planning: A Symbiotic Relationship

Weather plays an indispensable role in agriculture, and its influence extends to all aspects of agricultural planning. Effective agricultural planning relies heavily on understanding weather patterns, predicting future conditions, and adapting strategies to optimize production and minimize risks.

18.1.1. Understanding Weather's Influence on Agriculture

- **Crop Selection:** Different crops have specific climatic requirements, including temperature ranges, rainfall patterns, and sunlight duration. Historical weather data helps farmers determine the suitability of a particular region for growing certain crops.
- **Planting and Harvesting:** Weather patterns dictate optimal planting and harvesting times. For instance, farmers may delay planting if heavy rains are expected or expedite harvesting to avoid potential storms.
- **Irrigation Management:** Rainfall patterns and evapotranspiration rates influence irrigation needs. Weather forecasts help farmers schedule irrigation to ensure efficient water use and prevent water stress in crops.
- **Fertilizer and Pesticide Application:** Weather conditions affect the effectiveness of fertilizers and pesticides. Farmers need to consider rainfall, wind, and temperature when applying these inputs to maximize their efficacy and minimize environmental impact.
- **Disease and Pest Management:** Weather conditions can influence the outbreak and spread of plant diseases and pests. Farmers can use weather forecasts to anticipate potential problems and implement preventive measures.
- **Yield Prediction:** Weather patterns throughout the growing season significantly impact crop yields. Farmers use weather data to estimate

potential yields and make informed decisions about storage, marketing, and sales.

18.1.2. The Role of Weather Forecasting in Agricultural Planning

Weather forecasting is a critical tool for agricultural planning, providing farmers with valuable information to make timely and informed decisions.

- **Short-Range Forecasts (1-7 days):** These forecasts help farmers plan daily activities such as planting, harvesting, spraying, and irrigation.
- **Medium-Range Forecasts (7-14 days):** These forecasts are useful for making decisions about fertilizer application, pest and disease management, and water resource allocation.
- **Long-Range Forecasts (Seasonal and Monthly):** These forecasts help in crop selection, planting schedules, and overall farm management planning.
- **Extreme Weather Warnings:** Timely warnings about extreme weather events like droughts, floods, heatwaves, and frosts enable farmers to take preventive measures to protect their crops and livestock.

18.1.3. Integrating Weather Data into Agricultural Planning

Farmers and agricultural planners use various tools and technologies to integrate weather data into their decision-making processes:

- **Weather Stations:** On-farm or nearby weather stations provide real-time data on temperature, rainfall, humidity, wind speed, and other parameters.
- **Agrometeorological Models:** These models use weather data to simulate crop growth, predict yields, and assess the risk of pests and diseases.
- **Decision Support Systems (DSS):** These systems integrate weather data with other information, such as soil conditions and crop requirements, to provide farmers with recommendations on optimal management practices.
- **Remote Sensing and GIS:** Satellite imagery and geographic information systems provide valuable data on weather patterns, land use, and crop conditions, aiding in regional agricultural planning.
- **Mobile Apps and Online Platforms:** These tools provide farmers with easy access to weather forecasts, agricultural advisory, and other relevant information.

18.1.4. Challenges and Future Directions

Despite significant advances in weather forecasting and data integration, several challenges remain:

- **Forecast Accuracy:** Improving the accuracy of long-range forecasts and providing location-specific forecasts are crucial for effective agricultural planning.
- **Data Accessibility:** Ensuring that farmers, especially smallholder farmers in developing countries, have access to timely and reliable weather information is essential.
- **Climate Change:** The increasing frequency and intensity of extreme weather events due to climate change pose a significant challenge to agricultural planning.

Future directions in weather and agricultural planning include

- **Precision Agriculture:** Using технологии like sensors, drones, and machine learning to tailor agricultural practices to specific field conditions based on real-time weather data.
- **Climate-Smart Agriculture:** Adopting farming practices that enhance resilience to climate change, such as drought-resistant crops, water-efficient irrigation, and improved soil management.
- **Big Data and AI:** Leveraging big data analytics and artificial intelligence (AI) to improve weather forecasting, crop modeling, and decision support systems.

By effectively integrating weather data and forecasts into agricultural planning, farmers can optimize their operations, increase productivity, and ensure food security in a changing climate.

18.2. Crop Weather Relationships, Modeling and Decision Support Systems

18.2.1. Crop Weather Relationships

Crop weather relationships refer to how various weather elements (temperature, rainfall, humidity, solar radiation, wind, etc.) influence the growth, development, and yield of crops throughout their life cycle. These relationships are complex and vary depending on the crop species, its growth stage, and other environmental factors like soil type and nutrient availability. Understanding these relationships is crucial for:

- Optimizing planting and harvesting times.
- Predicting crop yields.

- Developing climate-smart agricultural practices.
- Assessing the impact of climate variability and change on agriculture.
- Designing effective irrigation and fertilization strategies.
- Forecasting pest and disease outbreaks influenced by weather conditions.

18.2.2. Studying Crop Weather Relationships

Researchers use various methods to study these relationships, including:

- **Field Experiments:** Conducting trials with different planting dates, irrigation regimes, and under varying weather conditions to observe crop responses.
- **Statistical Analysis:** Using historical weather and crop yield data to find correlations and build regression models.
- **Crop Simulation Models:** Employing computer models that simulate plant physiological processes based on weather inputs.

18.2.3. Crop Weather Modeling

Crop weather modeling involves using mathematical or computational models to simulate the growth and development of crops as a function of weather conditions, soil properties, and crop management practices. Types of crop models are:

- **Empirical/Statistical Models:** These models are based on statistical relationships observed between historical weather data and crop yields. They are often simpler but may not capture the underlying biological processes.
- **Mechanistic/Process-Based Models:** These models simulate the physiological processes of plant growth (e.g., photosynthesis, respiration, transpiration, nutrient uptake) based on weather inputs and plant characteristics. They are more complex but can provide insights into how crops respond to different environmental conditions.
- **Hybrid Models:** These models combine elements of both empirical and mechanistic approaches.

18.2.4. Applications of Crop Weather Models

- **Yield Forecasting:** Predicting crop yields before harvest.
- **Assessing Climate Change Impacts:** Simulating how future climate scenarios might affect crop production.
- **Optimizing Management Practices:** Determining the best planting dates, irrigation schedules, and fertilization rates under different weather conditions.

- **Risk Assessment:** Evaluating the likelihood of crop losses due to adverse weather events like droughts or heat waves.
- **Research:** Understanding plant responses to environmental stresses.

18.2.6. Decision Support Systems (DSS) in Agriculture

A Decision Support System (DSS) in agriculture is a computer-based tool that integrates data from various sources (weather, soil, crop models, market information, etc.) to provide farmers, agricultural advisors, and policymakers with information and recommendations to make better decisions

- **Key Components of Agricultural DSS**
  - **Database Management:** For storing and retrieving relevant data.
  - **Models:** Crop simulation models, statistical models, etc., to analyze data and generate predictions.
  - **User Interface:** To allow users to interact with the system easily.
  - **Knowledge Base:** Containing rules and information relevant to agricultural decision-making.
- **Examples of Applications of DSS in Agriculture**
  - **Crop Selection:** Recommending suitable crops based on climate, soil, and market demand.
  - **Planting Decisions:** Advising on optimal planting dates based on weather forecasts and crop phenology models.
  - **Irrigation Scheduling:** Providing recommendations on when and how much to irrigate based on soil moisture, weather, and crop water requirements.
  - **Nutrient Management:** Suggesting optimal fertilizer application rates based on soil tests and crop needs.
  - **Pest and Disease Management:** Forecasting the risk of outbreaks based on weather conditions and recommending timely control measures.
  - **Harvest Management:** Advising on the best time to harvest based on crop maturity and weather forecasts.
  - **Risk Management:** Helping farmers choose appropriate insurance or diversification strategies.
- **Benefits of Using DSS in Agriculture**
  - **Improved Efficiency:** Optimizing resource use (water, fertilizers, and pesticides).
  - **Increased Productivity:** Making more informed decisions that lead to higher yields.

- **Reduced Risks:** Helping to mitigate the impacts of adverse weather and other challenges.
- **Enhanced Sustainability:** Promoting environmentally sound practices.

In summary, understanding crop weather relationships allows for the development of crop weather models, which in turn form a crucial component of agricultural decision support systems. These integrated tools are increasingly important for enhancing agricultural productivity and sustainability in the face of climate variability and change.

18.3. Climate Change Impacts at National and Regional Level

18.3.1. National Level Impacts (India)

Climate change is already manifesting across India with significant consequences (Fig. 18.2):

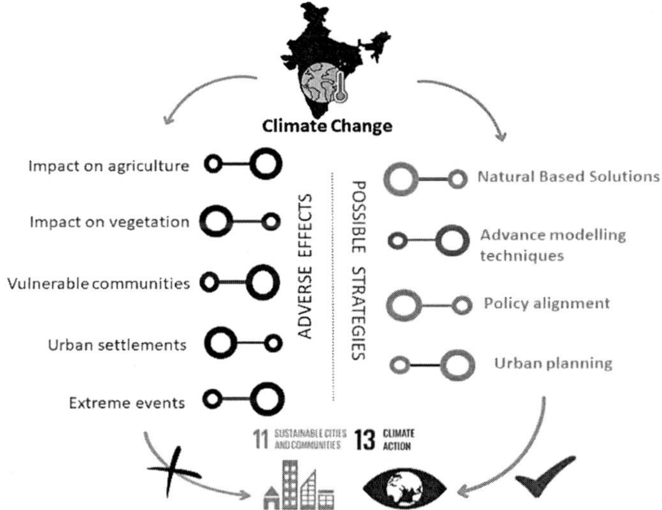

Fig. 18.2: Climate change impacts in India

Rising Temperatures and Heat Waves: India is experiencing a clear warming trend, with more frequent, intense, and longer heat waves. This poses severe risks to human health, agricultural labor productivity, and energy demand. For example, the 2022 heat wave impacted a large part of the country, affecting agriculture and human health.

- **Changes in Rainfall Patterns:** The Indian monsoon, crucial for agriculture and water resources, is becoming more unpredictable. There's evidence of a decline in monsoon rainfall since the 1950s, and future projections suggest more intense wet spells and drier dry spells. This increased variability can lead to both floods and droughts.

- **Increased Frequency and Intensity of Extreme Weather Events:** India is witnessing a rise in the frequency and intensity of cyclones, floods, and droughts. Coastal areas are particularly vulnerable to more intense cyclones and storm surges.
- **Glacier Melt and Impacts on Water Resources:** The Himalayan glaciers, the source of many major Indian rivers, are retreating. This threatens the long-term stability and reliability of river flows, impacting irrigation, hydropower, and the livelihoods of millions who depend on these rivers (Indus, Ganges, Brahmaputra). Initially, increased melt might lead to higher flows, followed by reduced flows in the long run.
- **Sea Level Rise and Coastal Degradation:** India has a long coastline vulnerable to sea-level rise, leading to saltwater intrusion, coastal erosion, and the submergence of low-lying areas. This impacts agriculture, groundwater quality, and can displace coastal communities. Cities like Mumbai and Kolkata are particularly at risk.
- **Impacts on Agriculture:** Changes in temperature and rainfall patterns, along with increased extreme events, are negatively affecting crop yields. Studies suggest potential declines in the yields of staple crops like rice and wheat. However, some crops like chickpea might see productivity increases.
- **Health Impacts:** Climate change is expected to exacerbate health risks through heat stress, changes in the spread of vector-borne diseases (like malaria and dengue), and increased air pollution. Extreme weather events also lead to injuries and displacement, impacting public health infrastructure.
- **Economic Impacts:** Climate change is projected to have substantial economic costs for India, with potential losses of GDP. Sectors like agriculture, water resources, and energy will be significantly affected.

18.3.2. Regional Level Impacts (India)

The impacts of climate change vary significantly across different regions of India due to diverse climate zones and geographical features:

- **Himalayan Region:** Facing accelerated glacier melt, changes in river flows, increased risk of glacial lake outbursts, and impacts on biodiversity. Agriculture and hydropower are particularly vulnerable.

- **Indo-Gangetic Plains:** Susceptible to increased heat stress affecting agriculture (especially wheat and rice), changes in monsoon patterns leading to floods and droughts, and air pollution exacerbated by meteorological conditions.
- **Northeast India:** Experiencing changes in rainfall patterns, potentially affecting the region's rich biodiversity and agricultural practices. Increased flooding and landslides are also risks.
- **Western Ghats:** Vulnerable to changes in rainfall, impacting the region's biodiversity hotspots and water resources. Extreme rainfall events and droughts can affect agriculture and ecosystems.
- **Coastal Regions:** Facing sea-level rise, increased intensity of cyclones and storm surges, saltwater intrusion affecting agriculture and drinking water, and threats to coastal ecosystems like mangroves.
- **Arid and Semi-Arid Regions (e.g., Rajasthan, parts of Deccan Plateau):** Likely to experience more intense and frequent droughts, increased heat stress, and challenges to water availability and agriculture.

19

Impact of Climate Change and Adaptation Strategies in Horticulture

Horticulture, the branch of agriculture that deals with the art, science, technology, and business of cultivating plants, plays a vital role in food security, nutrition, and economic development globally. This diverse sector encompasses a wide array of activities, broadly categorized into:

- **Pomology:** The cultivation of fruits and nuts.
- **Olericulture:** The cultivation of vegetables.
- **Floriculture:** The cultivation of ornamental flowering plants.
- **Landscape Horticulture:** The design, construction, and maintenance of landscapes.
- **Post-harvest Handling:** The management of horticultural produce after harvest to maintain quality and reduce losses.

Each of these components is intricately linked with weather conditions, making the horticulture sector highly vulnerable to climate change.

19.1. Role of Weather in Different Horticultural Activities

19.1.1. Pomology (Fruit and Nut Cultivation)

- **Temperature:** Temperature dictates the suitability of a region for specific fruit and nut crops (Fig. 19.1). It influences phenological stages like bud break, flowering, fruit set, fruit development, and maturity.

Fig. 19.1: Weather and horticultural activities

For example, temperate fruits like apples and peaches require specific chilling hours (exposure to low temperatures) for proper dormancy release and subsequent flowering. High temperatures during flowering can lead to poor pollination and fruit set, while extreme heat can cause sunburn and reduce fruit quality. The timing and intensity of heat units also determine the rate of fruit development and the harvest period.

- **Rainfall:** Adequate and timely rainfall is crucial for the growth and development of fruit and nut trees. Water stress, especially during critical stages like flowering and fruit development, can significantly reduce yields and quality. Conversely, excessive rainfall can lead to waterlogging, root diseases, and fruit cracking. The distribution of rainfall throughout the year is also important for perennial fruit crops.
- **Light:** Sunlight is essential for photosynthesis, which drives the growth and productivity of fruit and nut trees. The intensity and duration of light affect flowering, fruit development, color, and overall quality. Some fruit crops have specific light requirements for optimal growth and fruiting.
- **Humidity:** Humidity levels influence the incidence of pests and diseases in fruit and nut orchards. High humidity can favor the development of fungal diseases, while low humidity can increase water stress.
- **Wind:** While gentle wind can aid in pollination and reduce humidity, strong winds can damage blossoms, young fruits, and even tree branches.

19.1.2. Olericulture (Vegetable Cultivation)

- **Temperature:** Vegetables, being annual or biennial crops, are highly sensitive to temperature fluctuations throughout their growth cycle, from germination to harvesting. Temperature affects the rate of growth, flowering (in seed production), and fruit development. Extreme temperatures can cause bolting in some vegetables (premature flowering), reduce fruit set (e.g., in tomatoes), and affect the overall yield and quality.
- **Rainfall:** Similar to fruit crops, vegetables require adequate and timely water supply. The water requirements vary across different vegetable types and growth stages. Both drought and waterlogging can severely impact vegetable production.
- **Light:** Light intensity and duration play a crucial role in the vegetative growth and development of vegetables. Leafy vegetables require sufficient light for biomass production, while fruiting vegetables need it for flowering and fruit development. Photoperiod (day length) can also influence flowering in some vegetables.

- **Humidity:** Humidity affects transpiration rates and the susceptibility of vegetables to various fungal and bacterial diseases.

19.1.3. Floriculture (Ornamental Flower Cultivation)

- **Temperature:** Temperature is a critical factor in the growth, flowering, and quality of ornamental plants. It influences the timing of flowering, the size and color of blooms, and the overall plant health. Many floriculture crops have specific temperature requirements for different stages of their development.
- **Light:** Light intensity and photoperiod are particularly important in floriculture as they directly affect flowering initiation and development. Many ornamental plants are photoperiodic, meaning their flowering is triggered by specific day lengths. Light quality also influences flower color.
- **Water and Humidity:** Proper water management is essential for flower production. Both water stress and excessive moisture can be detrimental. Humidity levels affect the development of foliage and flowers and can also influence disease incidence.

19.1.4. Landscape Horticulture

- Weather primarily affects the selection and maintenance of plants used in landscaping. Temperature, rainfall, light availability, and wind patterns of a region dictate which plant species will thrive. Extreme weather events can also damage landscape installations.

19.1.5. Post-harvest Handling

- Temperature and humidity are the most critical weather-related factors affecting the shelf life and quality of harvested horticultural produce. Maintaining optimal temperature and humidity levels during storage and transportation is crucial to prevent spoilage and extend the marketability of fruits, vegetables, and flowers.

Given this strong dependence on weather, the predicted changes in climatic parameters pose significant challenges to the horticulture sector. The subsequent sections will delve into the specific impacts of climate change and the necessary adaptation strategies to ensure the resilience and sustainability of horticulture.

19.2. Impact of Climate Change on Pomology

Pomology, the science and practice of fruit and nut cultivation, is particularly vulnerable to the impacts of climate change due to the long-lived nature of perennial fruit trees and their specific requirements for temperature, water, and chilling. The observed and projected impacts include:

- **Shifts in Phenology:** Warmer temperatures can lead to earlier bud break, flowering, and fruit maturity. This can disrupt the synchronization between flowering and pollinator activity (e.g., bees), potentially reducing fruit set. Earlier flowering also increases the risk of damage from late spring frosts, as the sensitive flowers emerge before the last frost.
- **Altered Chilling Requirements:** Many temperate fruits (apples, cherries, peaches) require a certain number of chilling hours (exposure to low temperatures) during winter to break dormancy and flower properly. Warmer winters may result in insufficient chilling, leading to delayed or erratic flowering, reduced fruit set, and lower yields.
- **Heat Stress:** Increased frequency and intensity of heat waves can negatively affect fruit development, leading to sunburn, reduced fruit size, and lower quality. High temperatures during critical growth stages can also impair physiological processes within the trees.
- **Changes in Water Availability:** Altered rainfall patterns, including more intense rainfall events and prolonged droughts, can impact fruit production. Drought stress reduces tree vigor, fruit size, and yield, while excessive rainfall can lead to waterlogging, root diseases, and fruit cracking. Increased competition for water resources will also pose a challenge.
- **Increased Pest and Disease Pressure:** Warmer temperatures and altered humidity can favor the proliferation and spread of existing pests and diseases, as well as the emergence of new ones. This can lead to increased crop losses and the need for more intensive pest management.
- **Extreme Weather Events:** More frequent and intense extreme weather events like hailstorms, floods, and strong winds can directly damage fruit trees, blossoms, and fruits, leading to significant yield reductions and economic losses.
- **Changes in Fruit Quality:** Climate change can affect fruit quality parameters such as size, shape, color, sugar content, acidity, and nutritional value. For example, high temperatures can inhibit color development in some fruits.

19.2.1. Adaptation Strategies in Pomology

To mitigate the negative impacts of climate change on pomology and ensure sustainable fruit and nut production, various adaptation strategies are being explored and implemented:

- **Selection of Climate-Resilient Varieties:** This involves planting existing varieties or breeding new cultivars that are better adapted to warmer temperatures, reduced chilling, drought, or increased pest and disease pressure. For example, selecting low-chill varieties in regions with milder winters or drought-tolerant rootstocks in water-scarce areas.
- **Improved Irrigation and Water Management:** Implementing efficient irrigation systems like drip irrigation, rainwater harvesting, and soil moisture monitoring can help conserve water and mitigate the effects of drought.
- **Canopy Management and Shade Provision:** Techniques like pruning, training systems, and the use of shade nets can help regulate light and temperature within the orchard, reducing heat stress and sunburn on fruits.
- **Adjusted Orchard Management Practices:** This includes modifying planting times, fertilization schedules, and pest and disease management strategies to align with the changing climate and pest/disease dynamics. For instance, adjusting the timing of fertilizer application to avoid coinciding with heat waves.
- **Diversification of Crops and Varieties:** Planting a wider range of fruit and nut species or varieties with different climate tolerances can help buffer against losses from adverse weather events affecting specific crops.
- **Agroforestry Systems:** Integrating fruit trees with other plants (e.g., shade trees, cover crops) can improve soil health, water retention, and microclimate regulation.
- **Use of Protective Structures:** In some cases, using greenhouses or other protective structures can shield fruit crops from extreme weather events like hail and frost, and allow for better control of the growing environment.
- **Relocation of Orchards:** In the long term, some regions may become unsuitable for certain fruit crops, potentially necessitating the relocation of orchards to more climatically favorable areas. This is a complex strategy involving significant economic and social considerations.

- **Early Warning Systems:** Developing and utilizing climate forecasting and pest/disease prediction models can help growers make timely decisions to protect their crops.

The implementation of these adaptation strategies will require integrated efforts involving research, extension services, policymakers, and fruit growers to ensure the long-term sustainability and productivity of pomology in the face of a changing climate .

19.3. Impact of Climate Change on Olericulture

Olericulture, the cultivation of vegetables, is highly sensitive to climate change due to the annual or biennial nature of most vegetable crops and their often high water content (Fig. 19.2).

Fig. 19.2: Climate change impact on olericulture (Broccoli)

Climate change impacts on olericulture include

- **Temperature Effects**
  - **Heat Stress:** Increased temperatures can lead to reduced yields, poor quality (e.g., misshapen fruits, lower nutritional content), and physiological disorders like blossom-end rot in tomatoes. It can also cause premature bolting in crops like spinach and lettuce, making them unmarketable.
  - **Altered Growth Cycles:** Warmer temperatures may shorten the growing period for some vegetables, potentially leading to lower overall biomass accumulation and yield.
  - **Impact on Flowering and Fruiting:** High temperatures can interfere with pollination and fertilization in many vegetables (e.g., tomatoes, beans), leading to reduced fruit set.
- **Water Availability**
  - **Drought:** Changes in rainfall patterns, leading to more frequent and intense droughts, can cause severe water stress in vegetable crops,

which are generally shallow-rooted and require consistent moisture. This results in reduced growth, lower yields, and poor quality.

- **Flooding:** Conversely, increased heavy rainfall events can lead to waterlogging, causing root damage, promoting soil-borne diseases, and reducing oxygen availability to the roots, ultimately affecting plant survival and yield.

- **Increased Pest and Disease Pressure**
  - Warmer temperatures and altered humidity can create more favorable conditions for the growth and reproduction of many insect pests and fungal diseases, leading to increased infestation and disease incidence in vegetable crops. This necessitates more intensive pest and disease management.
- **Changes in Atmospheric CO_2**
  - While increased CO_2 can sometimes enhance photosynthesis, the associated increases in temperature and changes in water availability often negate these benefits in vegetable production. Furthermore, elevated CO_2 can sometimes reduce the nutritional quality of vegetables.
- **Extreme Weather Events**
  - Hailstorms, heat waves, and unseasonal frosts can cause direct physical damage to vegetable plants, leading to significant yield losses.

19.3.1. Adaptation Strategies in Olericulture

To address the challenges posed by climate change to vegetable cultivation in Hyderabad, Telangana, India, and elsewhere, various adaptation strategies can be implemented:

- **Developing and Using Climate-Resilient Varieties**
  - Breeding or selecting vegetable varieties that are tolerant to heat, drought, and flooding, as well as resistant to prevalent pests and diseases under changing climatic conditions.
  - Utilizing varieties with shorter growth cycles to avoid peak stress periods.
- **Improved Water Management:**
  - Implementing efficient irrigation techniques such as drip and micro-sprinkler irrigation to conserve water.

- Adopting rainwater harvesting and water-efficient soil management practices.
- Using mulching to reduce soil moisture loss and regulate soil temperature.

- **Protected Cultivation**
  - Utilizing greenhouses, shade nets, and plastic tunnels to create a more controlled environment, buffering crops against extreme temperatures, rainfall, and pests.
- **Adjusted Cropping Systems and Practices**
  - Altering planting and harvesting times to avoid the most stressful periods.
  - Practicing crop rotation to improve soil health and reduce pest and disease buildup.
  - Adopting intercropping systems to enhance resource utilization and resilience.
  - Promoting soil health through organic farming practices, which can improve water retention and nutrient availability.
- **Integrated Pest and Disease Management (IPM)**
  - Employing strategies that minimize reliance on chemical pesticides, focusing on biological control, resistant varieties, and cultural practices to manage pests and diseases effectively under changing climate conditions.
- **Agro-meteorological Information and Forecasting**
  - Utilizing weather forecasts and climate information to make informed decisions about planting, irrigation, and pest management.

Implementing these adaptation strategies requires collaborative efforts from researchers, extension services, policymakers, and vegetable growers to ensure the sustainability and productivity of olericulture in the face of a changing climate .

19.4. Impact of Climate Change on Floriculture

Floriculture, the cultivation of ornamental flowering plants, is also susceptible to the impacts of climate change in India, and globally. These impacts can affect various aspects of flower production:

- **Altered Phenology**
  - **Earlier Flowering:** Warmer temperatures can trigger earlier flowering in many species. This can disrupt market timing, especially for seasonal flowers.

- **Shorter Flowering Periods:** Some flowers may have a reduced blooming duration due to temperature changes.
- **Mismatched Synchronization:** Changes in flowering onset and duration can lead to mismatches with pollinator activity, potentially affecting seed production in seed-propagated varieties.

- **Temperature Extremes**
  - **Heat Stress:** High temperatures can lead to reduced flower size, altered flower color, wilting, and decreased overall quality. In some species, it can also reduce the number of flowering buds.
  - **Frost Damage:** Unexpected late frosts, especially with earlier flowering, can severely damage buds and blooms, leading to significant losses.
- **Changes in Water Availability**
  - **Drought:** Water stress can delay flowering, reduces flower size, and decrease the number of blooms. It can also impact the overall health and vigor of the plants.
  - **Excessive Rainfall/Humidity:** High humidity and excessive rainfall can promote fungal diseases and affect flower quality, leading to issues like petal blight.
- **Increased Pest and Disease Pressure**
  - Warmer temperatures and altered humidity can favor the proliferation and spread of various pests and diseases that affect ornamental plants.
- **Impact on Flower Quality**
  - Climate change can affect the fragrance of flowers due to changes in volatile organic compound emissions. It can also impact the intensity and consistency of flower color due to temperature effects on pigment production (Fig. 19.3). The post-harvest life of cut flowers can also be reduced under sub-optimal temperature conditions.

Fig. 19.3: Change in flower color under changing climate

- **Elevated CO_2**
  - While increased CO_2 can enhance photosynthesis and vegetative growth in some cases, it may not always translate to improved flower yield or quality. In some species, it can even lead to increased vegetative growth at the expense of reproductive output.

19.4.1. Adaptation Strategies in Floriculture

To mitigate the adverse effects of climate change on floriculture in Hyderabad, Telangana, India, and other regions, the following adaptation strategies can be considered

- **Selection of Climate-Resilient Varieties**
  - Choosing or breeding flower varieties that are more tolerant to heat, drought, or high humidity.
  - Selecting varieties with stable flower color and fragrance under varying temperature conditions.
- **Protected Cultivation**
  - Using greenhouses or shade houses to regulate temperature, humidity, and light intensity, providing a more controlled environment for flower production.
- **Efficient Water Management**
  - Implementing drip irrigation and other water-saving techniques.
  - Utilizing mulching to conserve soil moisture.
- **Adjusted Growing Schedules**
  - Altering planting and harvesting times to avoid periods of peak stress (e.g., extreme heat).
- **Integrated Pest and Disease Management (IPM)**
  - Adopting practices that help manage pests and diseases that may become more prevalent under changing climatic conditions.
- **Water Harvesting and Storage**
  - Collecting and storing rainwater for irrigation purposes.
- **Use of Biostimulants and Amendments**
  - Employing substances that can enhance plant tolerance to stress.
- **Diversification of Flower Crops**
  - Growing a wider range of flower species and varieties to reduce vulnerability to specific climate-related impacts.

Implementing these strategies will require research into the specific responses of different ornamental species to climate change and the development of best practices for growers in different regions.

19.5. Impact of Climate Change on Landscape Horticulture

Landscape horticulture, encompassing the design, installation, and maintenance of ornamental and functional landscapes, is influenced by climate change in several ways in Hyderabad, Telangana, India, and across the globe:

- **Plant Selection and Suitability**
  - **Shifting Hardiness Zones:** Warmer temperatures can lead to shifts in plant hardiness zones, making traditionally grown species less suitable for a region while opening opportunities for others. This necessitates a re-evaluation of plant palettes.
  - **Heat and Drought Stress:** Increased temperatures and altered rainfall patterns can lead to heat and drought stress in existing and newly planted landscapes, affecting plant health, growth rates, and survival.
  - **Changes in Native Plant Communities:** Climate change can alter the competitive dynamics within native plant communities, potentially leading to the decline of some species and the proliferation of others, including invasive ones.
- **Water Management**
  - **Increased Irrigation Needs:** Higher temperatures and reduced rainfall may increase the demand for irrigation in landscapes, putting a strain on water resources.
  - **Changes in Storm Patterns:** More intense rainfall events can lead to increased runoff and erosion in landscapes if not properly managed.
- **Pest and Disease Dynamics**
  - Warmer temperatures can expand the range and increase the activity of certain pests and diseases, requiring adjustments in plant health management practices in landscapes.
- **Extreme Weather Events**
  - More frequent and intense heat waves, storms, and other extreme weather events can cause direct damage to landscape plantings and structures.
- **Altered Growing Seasons:**
  - Changes in the timing of seasons can affect the optimal periods for planting, pruning, and other landscape maintenance activities.

19.5.1. Adaptation Strategies in Landscape Horticulture

To create resilient and sustainable landscapes in the face of climate change in India, and elsewhere, the following adaptation strategies are crucial:

- **Plant Selection for Resilience**
    - Choosing plant species and cultivars that are well-adapted to the projected future climate of the region, including heat and drought-tolerant varieties. Prioritizing native and locally adapted plants.
    - Selecting plants with lower water requirements (xeriscaping principles).
- **Water-Wise Landscaping**
    - Implementing water-efficient irrigation systems like drip irrigation.
    - Using mulch to retain soil moisture and reduce evaporation.
    - Designing landscapes to capture and infiltrate rainwater, reducing runoff.
- **Soil Health Management**
    - Improving soil structure and water-holding capacity through the addition of organic matter.
    - Reducing soil erosion through appropriate grading and planting.
- **Integrated Pest Management (IPM)**
    - Adopting IPM strategies to manage pests and diseases that may become more prevalent due to climate change.
- **Urban Heat Island Mitigation**
    - Increasing the use of green infrastructure (trees, green roofs, green walls) in urban landscapes to help mitigate the urban heat island effect.
- **Selection of Durable Materials**
    - Using hardscape materials that can withstand extreme weather conditions.
- **Flexible and Adaptive Design**
    - Designing landscapes that are more adaptable to changing conditions over time.

Implementing these adaptation strategies will help ensure that landscapes remain functional, aesthetically pleasing, and environmentally beneficial in a changing climate, urban areas in particular.

19.6. Impact of Climate Change on Post-Harvest Handling and Marketing

Climate change significantly impacts post-harvest handling and marketing of horticultural produce in India, and globally, primarily through:

- **Reduced Shelf Life:** Higher temperatures accelerate the ripening and senescence processes in fruits, vegetables, and flowers, leading to a shorter shelf life. This increases the risk of spoilage and losses during storage, transportation, and marketing.
- **Increased Pest and Disease Pressure:** Warmer temperatures and altered humidity can favor the growth and spread of post-harvest pests and diseases, leading to greater losses and reduced marketability.
- **Damage During Extreme Weather:** Extreme weather events like heat waves, floods, and storms can directly damage stored produce and disrupt transportation networks, leading to significant losses and market disruptions.
- **Changes in Produce Quality:** Pre-harvest climate conditions can affect the post-harvest quality of produce. For example, heat stress during growth can lead to physiological disorders that manifest during storage, affecting appearance, texture, and nutritional content.
- **Disruption of Supply Chains:** Extreme weather events can disrupt transportation, processing, and market access, affecting the smooth flow of horticultural produce from farm to consumer.

19.6.1. Adaptation Strategies for Post-Harvest Handling and Marketing

To minimize losses and ensure efficient marketing of horticultural produce under a changing climate, the following adaptation strategies are crucial:

- **Improved Cooling and Storage Facilities:** Investing in and expanding access to efficient cooling technologies (e.g., cold storage, evaporative cooling) at different points in the supply chain can significantly extend the shelf life of produce.
- **Enhanced Monitoring and Information Systems:** Implementing systems for monitoring temperature, humidity, and other environmental conditions during storage and transport can help in maintaining optimal conditions and reducing spoilage. Early warning systems for extreme weather events can also help in taking proactive measures.
- **Efficient Transportation and Logistics:** Optimizing transportation routes and using refrigerated transport where necessary can minimize losses during transit. Strengthening infrastructure to withstand extreme weather events is also important.

- **Modified Packaging:** Using appropriate packaging materials and techniques can help protect produce from physical damage and maintain quality during handling and transportation under varying climatic conditions.
- **Diversification of Markets and Processing:** Exploring alternative markets and increasing processing of horticultural produce into more stable forms can reduce dependence on fresh market sales and minimize losses due to shorter shelf life.
- **Capacity Building and Training:** Educating farmers, handlers, and marketers on best practices for post-harvest management under changing climate conditions is essential.
- **Research and Development:** Investing in research to develop post-harvest technologies and practices that are effective under warmer temperatures and more variable conditions is needed. This includes exploring innovative preservation techniques and packaging solutions.

By implementing these adaptation strategies, the post-harvest handling and marketing of horticultural produce in India, and other vulnerable regions can become more resilient to the impacts of climate change, ensuring food security and economic stability for stakeholders.

In conclusion, the diverse horticulture sector, encompassing fruits, vegetables, flowers, and landscapes, faces significant and multifaceted impacts from climate change globally. Alterations in plant phenology, heightened pest and disease pressure, evolving water availability, and the increasing frequency of extreme weather events collectively threaten the productivity, quality, and economic viability of this crucial industry.

- To ensure continued food security, economic stability, and the environmental and aesthetic benefits derived from horticulture worldwide, a critical imperative arises: the **intensification of sustainable horticultural cultivation**. This intensification must be fundamentally supported by an immediate and robust focus on **research**, **education**, and the development of **operational horticulture meteorology**.
- Accelerated research is essential to develop climate-resilient crop varieties, optimize resource management under shifting environmental conditions, and innovate post-harvest technologies that can mitigate increased spoilage. Concurrently, comprehensive educational programs are vital to empower growers and stakeholders with the knowledge and skills needed to adopt climate-smart horticultural practices effectively. Critically, the widespread integration of **operational horticulture meteorology** – providing tailored weather information, forecasts, and

agro-advisory services specific to horticultural activities – is paramount for enabling timely and informed decision-making, managing climate-related risks, and optimizing yields amidst increasingly unpredictable weather patterns.

- Delay in prioritizing and investing in intensive research, widespread education, and the establishment of robust operational horticulture meteorology will only exacerbate the adverse consequences of climate change on this vital sector globally. The urgency of the situation demands immediate and concerted action. A focused, adequately resourced, and collaborative global effort, beginning without further loss of time, is indispensable to safeguard the future of horticulture and the multitude of benefits it provides to societies and economies worldwide.

20

Impact of Climate Change on Sericulture

Sericulture, the art and science of silk production through silkworm rearing, is profoundly influenced by weather conditions. Optimal weather is not just beneficial; it's often a prerequisite for successful mulberry cultivation and healthy silkworm development. Deviations from ideal conditions can lead to decreased productivity, increased disease incidence, and ultimately, economic losses for sericulturists.

20.1. Weather's Influence on Mulberry Cultivation

Mulberry (*Morus* spp.) leaves are the sole food source for the most commercially important silkworm, *Bombyx mori*. Therefore, the health and productivity of mulberry plants are paramount. Weather parameters significantly affecting mulberry growth (Fig. 20.1) include:

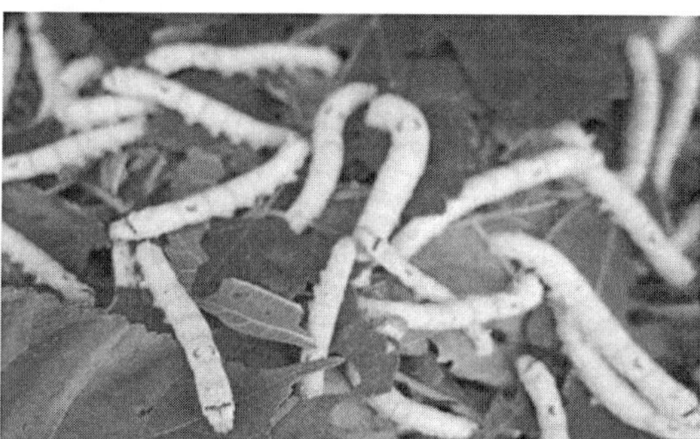

Fig. 20.1: Weather and mulberry silkworm rearing

Temperature: Mulberry thrives in a mean atmospheric temperature range of 24 to 28°C . While it can survive between 13.0 to 37.7°C, optimal growth and bud sprouting occur within the narrower ideal range. Temperatures outside this can stunt growth and leaf quality.

- **Rainfall:** An annual rainfall between 600 mm and 2500 mm is suitable for mulberry. In regions with lower rainfall, irrigation becomes essential. Consistent moisture is crucial for leaf yield and quality. Excessive rainfall can lead to waterlogging, harming the plants.

- **Humidity:** Relative humidity levels between 65% and 80% are generally favored for healthy mulberry growth.
- **Sunshine:** A daily sunshine duration of 5 to 12 hours supports optimal photosynthesis and overall development of mulberry plants.

20.1.1. Weather's Impact on Silkworm Rearing

Silkworms, being cold-blooded (ectothermic), are highly sensitive to environmental temperature and humidity. The ideal conditions for rearing *Bombyx mori* are:

- **Temperature:** The optimal temperature range is generally between 23 to 28°C although specific instars (larval stages) may have slightly different requirements. For instance, younger instars often prefer slightly higher temperatures 26 to 28°C while later instars do better in slightly cooler conditions 23 to 24°C.
    - High temperatures more than 30°C can accelerate larval development, potentially leading to lower silk yield and increased susceptibility to diseases.
    - Low temperatures less than 13°C can retard growth and prolong the larval period.
    - Stable temperatures are crucial, as significant fluctuations can negatively affect the silkworms' health and cocoon quality.
- **Humidity:** A relative humidity range of 70% to 85% is generally considered optimal, with younger instars requiring higher humidity (around 85%) compared to older ones (around 70%).
    - Low humidity can cause dehydration and affect silk quality.
    - High humidity can create an environment conducive to diseases.
- **Light and Air:** While not as critical as temperature and humidity, dim light (15-20 lux) is preferred over strong light or complete darkness. Good ventilation and fresh air are also essential to prevent the buildup of harmful gases like carbon dioxide and ammonia, which can negatively impact silkworm health.

20.2. Regions Practicing Sericulture in India

India is the second-largest producer of silk in the world, and sericulture is a significant agro-based industry providing livelihoods to millions, particularly in rural and semi-urban areas. The country produces all four major types of silk: Mulberry, Tasar, Eri, and Muga.

The predominant regions for sericulture in India, categorized by the type of silk primarily produced are:

20.2.1. Mulberry Silk

- **Karnataka:** Known as the "Silk Capital of India," Karnataka is the largest producer of mulberry silk, contributing significantly to the country's total output. Major sericulture districts include Ramanagara, Bengaluru, Mysuru, Kolar, and Chikkaballapura. The famous Mysore silk originates here.
- **Andhra Pradesh:** The second-largest producer of mulberry silk, with significant sericulture activities in districts like Anantapur, Chittoor, and Kadapa. Pochampally silk is a well-known variety from this region (though Pochampally is now in Telangana).
- **Tamil Nadu:** A significant producer of mulberry silk, with key centers in Kanchipuram, Arni, Salem, and Coimbatore. Kanchipuram silk sarees are world-famous.
- **West Bengal:** Another major mulberry silk-producing state, with concentrated activity in districts like Murshidabad, Birbhum, and Malda.
- **Jammu & Kashmir:** Also produces a notable quantity of mulberry silk, particularly in the Srinagar region.

20.2.2. Non-Mulberry Silks (Vanya Silks)

- **Tasar Silk:** Primarily produced in the states of **Jharkhand**, **Chhattisgarh**, and **Odisha**. Some production also occurs in **Maharashtra**, **West Bengal**, and **Andhra Pradesh**. These are mainly forest-based rearings involving the *Antheraea mylitta* silkworm that feeds on Asan and Arjun trees.
- **Eri Silk:** Predominantly produced in the **North-Eastern states**, with **Assam** being the largest producer. Other states in the region like Meghalaya, Nagaland, and Manipur also contribute. Eri silkworms (*Philosamia ricini*) mainly feed on castor leaves.
- **Muga Silk:** This golden-yellow silk is unique to **Assam**, produced by the *Antheraea assamensis* silkworm that feeds on Som and Soalu plants. Sualkuchi in Assam is a major center for Muga silk production.
- **Oak Tasar Silk:** Found in the sub-Himalayan belt, including states like **Manipur**, **Himachal Pradesh**, **Uttar Pradesh**, **Assam**, **Meghalaya**, and **Jammu & Kashmir**. The silkworm (*Antheraea proyeli*) feeds on oak leaves.

In summary, while **Karnataka** is the overall leader in silk production (mainly mulberry), sericulture is widespread across India, with different regions specializing in different types of silk. The **North-Eastern region** holds the unique distinction of producing all four commercial varieties of silk.

20.3. Regions Producing High Quality Silk

Identifying regions where high-quality silk is produced more profitably involves considering both the quality of the silk and the economic efficiency of its production. Here's a breakdown of regions known for high-quality silk and factors influencing profitability:

20.3.1. Regions Known for High-Quality Silk

- **Karnataka (especially Mysore):** Known for its high-quality mulberry silk, characterized by its softness, luster, and durability. The brand "Mysore Silk" has a strong reputation, which can command better prices. The adoption of improved bivoltine breeds in some areas contributes to higher quality.
- **Tamil Nadu (Kanchipuram):** Famous for Kanchipuram silk sarees, which are highly prized for their intricate gold zari work and rich colors. The skilled weaving tradition adds to the value and price.
- **Uttar Pradesh (Varanasi/Banaras):** Banarasi silk is renowned for its fine silk threads, intricate weaving patterns, and opulent designs, often featuring gold and silver brocade. These sarees are a luxury item.
- **Assam (Muga Silk):** Muga silk, with its natural golden luster and durability, is unique to Assam. Its exclusivity and distinctiveness can lead to higher value.

20.4. Factors Influencing Profitability

- **Silk Quality:** Higher quality silk (e.g., bivoltine mulberry silk with good grading) generally fetches better prices in the market, both domestically and internationally. Regions focusing on quality production can achieve higher profitability per unit.
- **Efficient Production Practices:** Adoption of modern techniques in mulberry cultivation (high-yielding varieties, better management), silkworm rearing (disease-free layings, scientific rearing practices), and reeling/spinning can reduce costs and improve yields, thus enhancing profitability.
- **Market Access and Branding:** Regions with well-established marketing channels, strong branding (like Mysore Silk or Kanchipuram

Silk), and access to premium markets (domestic and export) tend to be more profitable.

- **Government Support:** Schemes promoting quality silk production, providing subsidies, and supporting infrastructure can positively impact profitability.
- **Lower Production Costs:** Efficient use of resources, including labor, and locally available inputs can help reduce production costs.

20.5. Likely Regions for More Profitable High-Quality Silk Production

Considering both quality and potential for profitability, the following regions stand out:

- **Karnataka:** Due to its large scale of mulberry silk production, focus on quality (especially in certain pockets), established markets, and government support. Efforts to increase the production of high-grade bivoltine silk are likely to enhance profitability.
- **Tamil Nadu (Kanchipuram):** The premium pricing of Kanchipuram silk due to its craftsmanship and cultural significance contributes to higher profitability despite potentially higher production costs associated with intricate weaving.
- **Assam (Muga Silk):** The unique nature and cultural value of Muga silk allow it to command premium prices, potentially leading to higher profitability, although the scale of production might be smaller compared to mulberry silk regions.

It's important to note that profitability can vary within a state depending on specific practices, market linkages, and the type of silk produced. For instance, within Karnataka, areas focusing on higher grades of bivoltine silk for export or niche domestic markets might see greater profitability.

20.5.1. Regions Best Suitable for Sericulture

The best suitable sericulture regions, particularly in India, face significant challenges and potential shifts due to climate change. Here's an assessment:

- **Observed and Projected Impacts**
- **Temperature Increase:** Most climate models predict a rise in average temperatures in the major sericulture regions. This is concerning because silkworms are highly sensitive to temperature. Higher temperatures can lead to:
  - Accelerated larval development, potentially resulting in lower silk yield and quality.

  - Increased susceptibility to diseases in silkworms.
  - Heat stress in mulberry plants, affecting leaf quality and nutritional value for silkworms.
- **Altered Rainfall Patterns:** Changes in rainfall, including both increased intensity in some areas and droughts in others, can negatively impact sericulture:
  - Irregular or reduced rainfall can stress mulberry plants, affecting their growth and leaf production.
  - Increased heavy rainfall can lead to waterlogging, damaging mulberry plantations and potentially increasing disease risks for silkworms due to higher humidity.
- **Increased Frequency of Extreme Weather Events:** Regions may experience more heat waves, floods, and cyclones, which can directly damage mulberry crops and silkworm rearing infrastructure, leading to significant losses.
  - **Changes in Humidity:** Fluctuations in humidity can affect silkworm health and silk quality. Higher humidity can increase the risk of diseases, while lower humidity can lead to dehydration in silkworms.
  - **Pest and Disease Dynamics:** Warmer temperatures and altered humidity can influence the prevalence and distribution of pests and diseases affecting both mulberry and silkworms.
- **Regional Vulnerabilities To Climate Change**
- **Tropical Regions (e.g., Karnataka, Tamil Nadu, Andhra Pradesh, West Bengal, Assam):** These regions, which are major silk producers, are predicted to be severely impacted by a 2°C or greater average annual temperature rise. This could lead to significant reductions in mulberry leaf yield and silkworm productivity.
- **Temperate Regions (e.g., Jammu & Kashmir, sub-Himalayan areas):** While potentially less severely impacted than tropical regions, these areas may still experience changes affecting the timing of seasons and the suitability for specific silkworm breeds and mulberry varieties. Some studies suggest potential net revenue losses in temperate locations as well.
- **North-Eastern India:** This region, known for Eri and Muga silk, may see prolonged summers, impacting the prospects for Eri silkworm rearing. Changes in temperature and rainfall could also affect the specific host plants of Muga silkworms (Som and Soalu).

20.5.2. Adaptation Strategies

To mitigate the adverse effects of climate change, several adaptation strategies are being explored:

- Developing climate-resilient varieties of mulberry and silkworms.
- Improving water management and irrigation techniques.
- Adopting integrated pest and disease management.
- Using controlled environment rearing houses to buffer against external temperature and humidity fluctuations.
- Adjusting rearing schedules to align with changing seasonal patterns.

In conclusion, the assessment of climate change in the best sericulture regions of India indicates a significant threat to this vital agricultural industry. Rising temperatures and altered rainfall patterns pose considerable risks to both mulberry cultivation and silkworm rearing. Implementing adaptive strategies and investing in research to develop climate-resilient practices will be crucial for the sustainability of sericulture in the future.

20.5.3. The Growing Threat of Climate Change

Climate change introduces significant challenges to sericulture through:

- **Increased Temperatures:** Higher average temperatures and more frequent heat waves can stress both mulberry plants and silkworms, potentially reducing leaf quality and silk yield.
- **Altered Rainfall Patterns:** Changes in rainfall, including droughts and floods, can directly impact mulberry cultivation and indirectly affect silkworm rearing.
- **Extreme Weather Events:** More frequent and intense extreme weather events can damage mulberry plantations and disrupt silkworm rearing environments.
- **Increased Pest and Disease Incidence:** Changing climatic conditions can favor the proliferation of pests and diseases affecting both mulberry and silkworms.

20.5.4. Adaptation and Mitigation

To ensure the sustainability of sericulture in the face of changing weather patterns, various adaptation and mitigation strategies are being explored, including:

- Breeding climate-resilient varieties of mulberry and silkworms.
- Improving irrigation and water management practices.

- Adopting integrated pest and disease management strategies.
- Utilizing controlled rearing environments to buffer against external weather fluctuations.

In conclusion, the success of sericulture is deeply intertwined with favorable weather conditions. Understanding these relationships and adapting to the challenges posed by climate change is crucial for the future of this important agro-industry.

20.6. Importance of Sericulture in Indian Economy

Sericulture, the cultivation of silkworms for silk production, holds significant importance for the Indian economy due to several factors:

- **Livelihood Generation:** It is a labor-intensive, agro-based cottage industry providing employment to millions, particularly in rural and semi-urban areas. It supports marginal farmers, landless laborers, and artisans, contributing significantly to rural income and poverty alleviation.
- **Women Empowerment:** A significant portion of the workforce in sericulture, particularly in activities like silkworm rearing and post-cocoon processing, comprises women, thus playing a vital role in their economic empowerment.
- **Agricultural Linkage:** Sericulture is closely linked to agriculture, with mulberry cultivation being the primary activity. This integration provides farmers with an additional income source and promotes diversification in agriculture.
- **Foreign Exchange Earnings:** India is a major silk producer and consumer globally. The export of silk and silk goods contributes to the country's foreign exchange earnings. In 2023-24, the export of silk and silk goods was valued at □2,027.56 crores.
- **Regional Development:** Sericulture is practiced across various states, each often specializing in a particular type of silk (Mulberry, Tasar, Eri, Muga). This helps in the economic development of diverse regions, including the North-Eastern states known for non-mulberry silks.
- **Low Investment and High Returns:** Compared to some other industries, sericulture often requires lower initial investment and has a relatively short gestation period, offering quicker returns, which is beneficial for small and marginal farmers.
- **Eco-Friendly Activity:** Mulberry cultivation helps in soil conservation, and silkworm waste can be used as manure, making it a relatively eco-friendly agricultural practice.

20.7. Need for Intensive Research in Sericulture

Despite its importance, the sericulture industry faces several challenges that necessitate intensive research for its sustained growth (Fig. 20.2) and profitability:

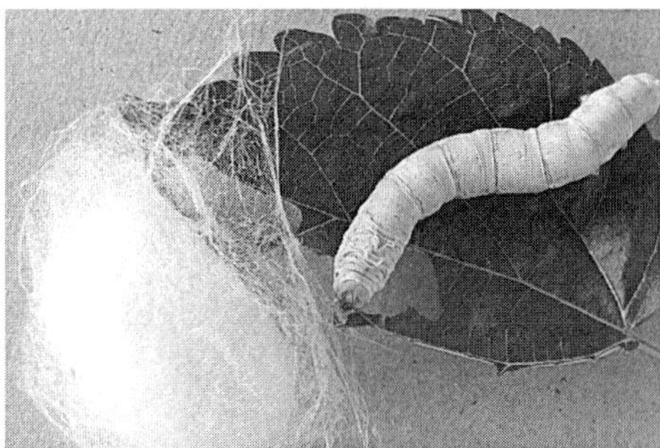

Fig. 20.2: Intensive research in sericulture

Climate Change Impacts: As discussed earlier, climate change poses a significant threat through temperature variations, altered rainfall, and increased extreme weather events, affecting both mulberry and silkworm health and productivity. Research is needed to develop climate-resilient breeds of mulberry and silkworms and adaptive management practices.

- **Disease Management:** Silkworms are susceptible to various diseases that can cause significant crop losses. Intensive research is required to develop effective disease prevention and management strategies, including the use of biotechnology for disease resistance.
- **Improving Silk Quality and Productivity:** There's a continuous need to enhance the quality (luster, strength, texture) and yield of silk. Research in silkworm genetics and breeding, as well as advancements in rearing technologies, can contribute to this. For instance, research into bivoltine breeds known for higher quality silk needs to be intensified and adapted to local conditions.
- **Technological Advancements:** Adopting modern technologies in various stages of sericulture, from mulberry cultivation to silk reeling and processing, can improve efficiency and reduce costs. Research in automation, climate-controlled rearing, and digital monitoring is crucial.
- **Value Addition and Diversification:** Research into innovative uses of silk beyond textiles, such as in biomaterials for medical applications or cosmetics, and the utilization of sericulture by-products (e.g., silkworm pupae, mulberry waste) can enhance the economic viability of the industry.

- **Pest Management in Mulberry:** Pests can significantly damage mulberry crops, affecting the quality and quantity of leaves available for silkworms. Research into integrated pest management strategies that are effective and environmentally friendly is essential.
- **Post-Cocoon Technology:** Improving reeling, spinning, and weaving technologies can lead to better quality yarn and fabrics, fetching higher market prices. Research in these areas can enhance the value chain of silk production.

Intensive and focused research, supported by government and other institutions, is vital to address these challenges and harness the full potential of sericulture for the Indian economy, ensuring its sustainability and the prosperity of the millions involved in it.

21

Agricultural Meteorology: An Essential Associate for Food and Nutrition Security

In an era defined by increasing climate variability and the urgent need for sustainable food production, the science of Agricultural Meteorology has risen to unprecedented importance. This final chapter, "Agricultural Meteorology: A Science for Global and National Food Security (in the context of climate change)," synthesizes the principles and applications we have explored, highlighting their critical role in ensuring food security at both global and national levels, particularly within the challenging framework of a changing climate.

We have seen how weather and climate intricately influence every aspect of agriculture, from planting to harvest, and how extreme events can decimate livelihoods and threaten food supplies.1 Now, we will focus on how the interdisciplinary field of agricultural meteorology – the study of the interactions between meteorological and hydrological factors and agriculture in the widest sense – provides the knowledge and tools necessary to navigate these challenges.

This chapter will underscore how understanding and applying agro-meteorological principles, including crop-weather relationships, climate risk assessment, and the use of decision support systems, are essential for developing climate-resilient agricultural practices. We will explore how this science aids in optimizing resource management, predicting and mitigating the impacts of weather-related hazards, and ultimately, in building sustainable and secure food systems for the future, both globally and specifically within the Indian context.

Join us as we conclude our exploration by examining the pivotal role of agricultural meteorology in safeguarding the food security of nations and the world in the face of a rapidly evolving climate.

21.1. Development of Agricultural Meteorology: Priorities

It's crucial to prioritize education, research, agromet advisory services, and capacity building at the regional level to effectively address the unique challenges posed by climate change to agriculture. Here's a detailed look at the needs for different stakeholders:

21.1.1. Farmers/Growers

- **Education**
  - **Need:** Farmers require practical, location-specific education on climate change impacts, sustainable farming practices, water management, and pest/disease control under changing weather patterns (Fig. 21.1).
  - **Detail:** This should include training on:
    - Understanding local climate change projections and their effects on crops.
    - Implementing climate-resilient agricultural techniques (e.g., drought-resistant varieties, intercropping, conservation tillage).
    - Efficient irrigation methods and water harvesting.
    - Integrated pest and disease management strategies adapted to changing conditions.
    - Using technology (e.g., mobile apps, weather alerts) for informed decision-making.

Fig. 21.1: Weather and farming

Research

- **Need:** Participatory research that involves farmers in identifying problems and testing solutions relevant to their specific regional contexts.
- **Detail:** This includes:
  - On-farm trials to evaluate the performance of different crop varieties and management practices under local climate conditions.
  - Studies on the impact of climate variability on traditional farming systems and indigenous knowledge.

- Development of climate-smart agricultural technologies that are affordable and accessible to smallholder farmers.

- **Agrometeorological Advisory Services**
  - **Need:** Timely, accurate, and location-specific agromet advisories delivered through accessible channels.
  - **Detail:** This involves
    - Providing weather forecasts (short-term, medium-range, and seasonal) tailored to specific agricultural zones within the region.
    - Developing advisories that recommend specific actions (e.g., when to plant, irrigate, apply inputs, harvest) based on weather forecasts and crop conditions.
    - Disseminating information through multiple channels, including:
    - Mobile phones (SMS, voice messages, apps)
    - Local radio and television
    - Community meetings and extension services
    - Digital platforms and social media
    - Ensuring advisories are in local languages and easy to understand.
- **Capacity Building**
  - **Need:** Strengthening farmers' ability to access and use information, adopt new technologies, and adapt to climate change.
  - **Detail:** This includes
    - Training on how to interpret and apply agromet advisories.
    - Facilitating farmer-to-farmer learning and knowledge sharing.
    - Supporting the formation of farmer groups and cooperatives to enhance collective action.
    - Providing access to credit, inputs, and markets to enable the adoption of climate-resilient practices.

21.1.2. Extension Workers

- **Education**
  - **Need:** Equipping extension workers with the latest knowledge and skills in agrometeorology, climate change adaptation, and sustainable agriculture.
  - **Detail:** This should cover:
    - Advanced training in interpreting weather and climate data and generating agromet advisories.

    - Understanding the regional impacts of climate change on agriculture.
    - Promoting climate-smart agricultural practices and technologies.
    - Effective communication and training methodologies to reach diverse farmer groups.
    - Use of ICT tools for information dissemination and data collection.
- **Research**
  - **Need:** Involving extension workers in applied research to ensure that technologies and practices are relevant to local needs.
  - **Detail:** This includes:
    - Participating in on-farm trials and demonstrations to evaluate the effectiveness of climate-resilient practices.
    - Collecting feedback from farmers on the usefulness of agromet advisories and identifying areas for improvement.
    - Documenting and sharing successful adaptation strategies and indigenous knowledge.
- **Agrometeorological Advisory Services**
  - **Need:** Providing extension workers with the tools and resources to deliver effective agromet advisory services.
  - **Detail:** This involves:
    - Access to reliable weather and climate information and decision support systems.
    - Training on how to tailor advisories to specific crops, locations, and farmer needs.
    - Developing communication strategies to reach a large number of farmers in a timely manner.
- **Capacity Building**
  - **Need:** Enhancing the capacity of extension services to support farmers in adapting to climate change.
  - **Detail:** This includes:
    - Strengthening institutional capacity through training, infrastructure development, and resource allocation.
    - Promoting collaboration between extension agencies, research institutions, and other stakeholders.
    - Developing innovative extension approaches (e.g., farmer field schools, participatory extension)

21.1.3. Policymakers and Government Officials

- **Education**
  - **Need:** Sensitizing policymakers to the importance of agrometeorology and the need for climate-resilient agricultural policies.
  - **Detail:** This includes:
    - Providing information on the economic and social impacts of climate change on agriculture in the region.
    - Training on how to integrate agrometeorological information into policy planning and decision-making.
    - Workshops on best practices in climate change adaptation and mitigation in the agricultural sector.
- **Research**
  - **Need:** Providing policymakers with scientific evidence to inform the development of effective climate change adaptation and mitigation policies.
  - **Detail:** This involves:
    - Supporting research on the long-term impacts of climate change on regional agriculture.
    - Developing models and tools to assess the vulnerability of agricultural systems to climate risks.
    - Evaluating the effectiveness of different policy interventions.
- **Agrometeorological Advisory Services**
  - **Need:** Utilizing agromet information to inform policy decisions related to agriculture, disaster management, and resource allocation.
  - **Detail:** This includes:
    - Using seasonal forecasts to plan for drought or flood preparedness.
    - Integrating agromet data into early warning systems for extreme weather events.
    - Developing policies to promote the adoption of climate-resilient agricultural practices.
- **Capacity Building**
  - **Need:** Strengthening the capacity of government institutions to address the challenges of climate change in the agricultural sector.
  - **Detail:** This involves:
    - Establishing inter-departmental coordination mechanisms.

- Providing training to government officials on climate change adaptation and mitigation.
- Allocating resources for the development and implementation of climate-resilient agricultural programs.

21.1.4. Researchers and Scientists

- **Education**
  - **Need:** Continuous learning and specialization in advanced agrometeorology, climate modeling, and related fields.
  - **Detail:** This includes:
    - Supporting postgraduate programs and research fellowships in agrometeorology.
    - Organizing specialized workshops and conferences on emerging topics.
    - Promoting interdisciplinary collaboration with other fields (e.g., agronomy, ecology, economics).
- **Research**
  - **Need:** Conducting cutting-edge research to improve our understanding of crop-weather relationships, climate change impacts, and adaptation strategies (Fig. 21.2).

Fig. 21.2: Weather research for farm advisories

  - **Detail: This involves**
    - Developing and validating crop simulation models for regional applications.
    - Improving the accuracy and resolution of weather and climate forecasts.
    - Assessing the vulnerability of different agricultural systems to climate change.

  - Developing new climate-resilient crop varieties and farming technologies.
  - Studying the socio-economic impacts of climate change on agriculture.
- **Agrometeorological Advisory Services**
  - **Need:** Providing the scientific basis for the development of effective agromet advisory services.
  - **Detail:** This includes:
    - Developing methodologies for translating weather forecasts into actionable agricultural advice.
    - Evaluating the impact and effectiveness of agromet advisories.
    - Developing decision support tools for farmers and extension workers.
- **Capacity Building**
  - **Need:** Building a strong scientific community with the expertise to address the challenges of climate change in agriculture.
  - **Detail:** This involves:
    - Investing in research infrastructure and facilities.
    - Promoting collaboration and networking among researchers at the regional, national, and international levels.
    - Providing funding opportunities for innovative research projects.

By addressing these needs for each stakeholder group at the regional level, we can build more resilient agricultural systems, enhance food security, and improve the livelihoods of those most vulnerable to the impacts of climate change.

21.2. Weather / Meteorology - An Essential Associate

Weather / meteorology is an **essential associate of all production systems on the planet Earth** because weather and climate fundamentally influence every sector that depends on natural resources, energy, infrastructure, and human activity. Meteorology integrates with and supports various production systems (Fig. 21.3) *viz.*:

Fig. 21.3: Agrometeorology – an essential associate

- **Agriculture and Food Systems**
  - **Weather forecasting** informs planting, irrigation, pest control, and harvesting.
  - **Climate data** supports crop selection and long-term agricultural planning.
  - Helps in **drought and flood risk management**, directly affecting crop yields and food security.
- **Water Resources and Hydropower**
  - Rainfall patterns and evaporation rates, derived from meteorological data, are crucial for **water reservoir management**.
  - In hydropower production, **river flow forecasts** ensure optimal power generation and dam safety.
- **Energy Sector**
  - Solar, wind, and hydro energy systems depend on **weather variability**.
  - Forecasts guide **load balancing** and energy distribution.
  - **Extreme weather alerts** help protect energy infrastructure.
- **Transportation and Logistics**
  - Aviation, shipping, and road transport rely on meteorology for **safety, routing, and fuel efficiency**.
  - Weather disruptions can impact supply chains and delivery schedules.
- **Health Systems**
  - Meteorology supports early warning for **heatwaves, cold spells, and air quality** concerns.
  - It helps manage the spread of **climate-sensitive diseases** like malaria or dengue.

- **Industry and Manufacturing**
  - Weather impacts **raw material supply, factory operations**, and even labor availability.
  - Industries that rely on outdoor processes (e.g., construction, mining) use forecasts to plan work and protect assets.
- **Disaster Risk Reduction:**
  - **Early warning systems** for storms, floods, droughts, etc., help prevent losses across production systems.
  - Supports **resilience planning** in climate-vulnerable areas.
- **Environmental Monitoring and Ecosystem Services:**
  - Meteorology is essential in tracking **climate change**, biodiversity impacts, and **natural resource management**.
  - Forests, fisheries, and other ecological production systems are weather-dependent.

Meteorology acts as the **invisible thread** connecting natural cycles with human production systems. It enables **anticipation, adaptation, and resilience**, making it a critical enabler of **sustainable development and economic productivity all along**.

Chapter wise 50 Multiple Choice Questions (MCQs)

Ch. 1. Introduction to Meteorology and Agricultural Meteorology

1. Agricultural meteorology is the study of
   a) Soil biology
   b) Weather and climate in relation to crops
   c) Ocean tides
   d) Livestock breeding
2. The term 'meteorology' originates from:
   a) Latin b) Greek
   c) Sanskrit d) Arabic
3. Who is considered the father of meteorology?
   a) Newton b) Aristotle
   c) Galileo d) Einstein
4. Agrometeorology deals with the interaction between
   a) Farmers and crops
   b) Plants and soil
   c) Meteorological parameters and agricultural production
   d) Livestock and climate
5. Which of the following best describes agrometeorology?
   a) Physics of atmosphere
   b) Applied science combining meteorology and agriculture
   c) Study of rocks
   d) Study of animal diseases
6. Agricultural meteorology is useful for
   a) Land reforms b) Fertilizer manufacture
   c) Weather-based farming decisions d) Genetic engineering

7. Which operation is least affected by weather?
   a) Sowing b) Harvesting
   c) Weeding d) Fertilizer application
8. Weather forecasts help farmers in:
   a) Post-harvest marketing
   b) Selection of seed variety
   c) Timely operations like irrigation and pesticide spraying
   d) Farm mechanization
9. Climatic factors influence:
   a) Soil formation only
   b) Crop duration, yield and quality
   c) Only seed germination
   d) None of these
10. Agrometeorological services provide advisories mainly for:
    a) Policymakers b) Extension workers
    c) Farmers d) Meteorologists
11. Which of the following is not a component of agricultural meteorology?
    a) Soil moisture b) Wind
    c) Sunshine duration d) Seed viability
12. Meteorological factors include:
    a) Temperature b) Rainfall
    c) Humidity d) All of the above
13. Which branch of meteorology deals with climate and agriculture?
    a) Synoptic meteorology b) Climatology
    c) Agricultural meteorology d) Oceanography
14. The instrument used to measure humidity is:
    a) Barometer b) Hygrometer
    c) Anemometer d) Pyranometer
15. Which instrument measures solar radiation?
    a) Pyranometer b) Wind vane
    c) Rain gauge d) Soil thermometer

16. Which is a key scope of agrometeorology?
    a) Cloud seeding
    b) Pesticide formulation
    c) Fertilizer recommendation
    d) Weather forecasting for agriculture
17. Weather has the greatest impact on
    a) Machinery efficiency
    b) Input cost
    c) Crop productivity
    d) Market pricing
18. Extreme weather conditions like frost or drought can lead to
    a) Better yields
    b) Reduced crop growth
    c) Increased transpiration
    d) None of the above
19. Heat units are also referred to as
    a) Thermal indexes
    b) Crop coefficients
    c) Growing degree days (GDD)
    d) Wind chill factor
20. Wind breaks and shelterbelts help in
    a) Increasing rainfall
    b) Enhancing soil fertility
    c) Reducing wind speed & erosion
    d) Soil acidification
21. IMD (India Meteorological Department) was established in
    a) 1835
    b) 1875
    c) 1947
    d) 1950
22. The headquarters of IMD is located at
    a) Mumbai
    b) New Delhi
    c) Pune
    d) Chennai
23. Agrometeorological Field Units (AMFUs) are coordinated by:
    a) ICAR
    b) IMD
    c) Ministry of Agriculture
    d) State governments
24. Which organization provides weather-based agro-advisories in India?
    a) IARI
    b) ICAR
    c) IMD
    d) ISRO
25. First agrometeorological observatory in India was established at:
    a) Pune
    b) New Delhi
    c) Bangalore
    d) Hyderabad

26. Weather is the atmospheric condition of a place
    a) Over a long period b) In the past
    c) At a specific time d) Globally
27. Climate is generally considered as the average weather over
    a) 1 year b) 5 years
    c) 10 years d) 30 years
28. Which of the following is a climatic element?
    a) Cloud type b) Humidity
    c) Lightning d) Thunder
29. Which statement is true?
    a) Climate changes every hour
    b) Weather is a long-term phenomenon
    c) Weather varies daily, climate is average over time
    d) Climate and weather are synonyms
30. Long-term variations in temperature and rainfall are part of:
    a) Climate b) Weather
    c) Season d) Daily forecast
31. Which discipline helps agrometeorology understand crop-water needs?
    a) Genetics b) Soil science
    c) Hydrology d) Entomology
32. Meteorological data help in:
    a) Soil reclamation b) Crop zoning
    c) Forest fire prevention d) Pest management
33. Which is a non-meteorological factor affecting crop yield?
    a) Soil type b) Rainfall
    c) Sunshine d) Temperature
34. Satellite-based agromet observations support:
    a) Crop insurance b) Fertilizer purchase
    c) Market price analysis d) Pest resistance breeding
35. Weather influences livestock by:
    a) Affecting feed intake b) Improving immunity
    c) Reducing disease resistance d) Altering genetic makeup

36. Climate-smart agriculture promotes
    a) Increased fossil fuel use
    b) Unsustainable practices
    c) Resilience to climate variability
    d) High greenhouse emissions
37. Weather-based insurance schemes are aimed to
    a) Reduce bank interest
    b) Subsidize fertilizer
    c) Protect farmers from crop losses due to weather
    d) Increase storage costs
38. Which technology helps in large-scale weather data collection?
    a) Drone spraying
    b) Remote sensing
    c) Manual records
    d) Soil testing
39. Digital tools in agrometeorology enable
    a) Increase in tillage operations
    b) Delay in sowing
    c) Real-time weather updates
    d) Manual forecasting
40. Which is not a scope of agrometeorology?
    a) Crop yield prediction
    b) Disease forecasting
    c) Soil classification
    d) Pest forecasting
41. Farmers use weather advisories to decide on
    a) Land ownership
    b) Crop sowing dates
    c) Legal issues
    d) Farm accounting
42. Weather advisory services are most critical during
    a) Off-season
    b) Crop growth periods
    c) Post-harvest
    d) Storage
43. Farmers receive agro-meteorological advisories through
    a) Radio and mobile phones
    b) Soil labs
    c) Machinery tools
    d) Farm inputs
44. Timely weather information improves
    a) Farm mechanization
    b) Crop productivity and risk reduction
    c) Export opportunities
    d) Pesticide residues
45. Which is not included in agro-advisory bulletin?
    a) Weather forecast
    b) Cultural operations
    c) Legal notice
    d) Pest control tips

46. The biggest challenge in agrometeorology is
    a) Pesticide resistance
    b) Precision in local forecasts
    c) High temperature
    d) Fertilizer cost
47. Uncertainty in rainfall forecasting affects
    a) Fertilizer availability
    b) Pesticide formulation
    c) Water management & sowing
    d) Tractor hire
48. Lack of farmer awareness limits the use of
    a) Machinery
    b) Agro-meteorological services
    c) Fertilizer bags
    d) Extension services
49. High-resolution agromet models are based on
    a) Historical yield data
    b) Weather observations and simulations
    c) Marketing trends
    d) Soil nutrient levels
50. Integration of AI in agrometeorology helps in
    a) Irrigation design
    b) Forecast accuracy and decision support
    c) Manual weather logging
    d) Fertilizer production

Answer Key

1	b	2	b	3	b	4	c	5	b	6	c	7	c
8	c	9	b	10	c	11	d	12	d	13	c	14	b
15	a	16	d	17	c	18	b	19	c	20	c	21	b
22	b	23	b	24	c	25	a	26	c	27	d	28	b
29	c	30	a	31	c	32	b	33	a	34	a	35	a
36	c	37	c	38	b	39	c	40	c	41	b	42	b
43	a	44	b	45	c	46	b	47	c	48	b	49	b
50	b												

Ch. 2. Earth‘s Atmosphere, Weather and Climate

1. The most abundant gas in the Earth’s atmosphere is
   a) Oxygen b) Carbon dioxide
   c) Nitrogen d) Argon
2. Nitrogen constitutes approximately what percentage of dry air?
   a) 21% b) 78%
   c) 0.03% d) 50%
3. Oxygen makes up around ___% of the Earth’s atmosphere by volume
   a) 25% b) 78%
   c) 21% d) 10%
4. Which gas acts as a greenhouse gas?
   a) Nitrogen b) Oxygen
   c) Carbon dioxide d) Argon
5. Water vapor in the atmosphere varies between
   a) 0-1% b) 0-4%
   c) 1-10% d) 10-20%
6. Carbon dioxide concentration in the atmosphere is approximately
   a) 0.03% b) 1%
   c) 10% d) 3%
7. Argon in the atmosphere is about
   a) 5% b) 1.5%
   c) 0.93% d) 0.01%
8. Which gas is present in trace amounts but important for ozone formation?
   a) Hydrogen b) Methane
   c) Nitrous oxide d) Ozone
9. Which of the following is a variable component of the atmosphere?
   a) Argon b) Nitrogen
   c) Water vapor d) Oxygen

11. The presence of dust particles in the atmosphere helps in
   a) Absorbing radiation b) Cloud formation
   c) Scattering sunlight d) All of the above
12. The atmosphere extends up to about
   a) 10 km b) 50 km
   c) 500 km d) 10,000 km

13. The lowest layer of the atmosphere is called
    a) Stratosphere b) Troposphere
    c) Mesosphere d) Thermosphere
14. The layer where weather phenomena occur is
    a) Thermosphere b) Mesosphere
    c) Troposphere d) Exosphere
15. The average height of the troposphere at the equator is around
    a) 8 km b) 11 km
    c) 16-18 km d) 25 km
16. The boundary between the troposphere and stratosphere is called
    a) Stratopause b) Tropopause
    c) Mesopause d) Exopause
17. The temperature in the stratosphere
    a) Decreases with height b) Increases with height
    c) Remains constant d) Varies irregularly
18. The ozone layer is mainly found in
    a) Troposphere b) Stratosphere
    c) Mesosphere d) Exosphere
19. The coldest layer of the atmosphere is
    a) Troposphere b) Mesosphere
    c) Stratosphere d) Thermosphere
20. Auroras occur in which layer of the atmosphere?
    a) Troposphere b) Mesosphere
    c) Thermosphere d) Stratosphere
21. The outermost layer of the atmosphere is
    a) Exosphere b) Thermosphere
    c) Mesosphere d) Troposphere
21. Stratosphere is ideal for flying jet aircraft because
    a) It is turbulent
    b) It contains clouds
    c) It is stable and free from weather disturbances
    d) It has strong winds

22. Temperature decreases with height in the
    a) Troposphere and Mesosphere
    b) Stratosphere & Thermosphere
    c) Exosphere only
    d) All layers
23. The Mesosphere extends roughly from
    a) 0–10 km
    b) 10–50 km
    c) 50–80 km
    d) 80–500 km
24. The Thermosphere contains
    a) Ionosphere
    b) Ozone maximum
    c) Jet streams
    d) Rain clouds
25. Which layer helps in radio communication?
    a) Troposphere
    b) Mesosphere
    c) Ionosphere (in Thermosphere)
    d) Stratosphere
26. The rate at which temperature decreases with height in the troposphere is called
    a) Lapse rate
    b) Saturation rate
    c) Dew rate
    d) Radiative cooling
27. The normal lapse rate is
    a) 2°C per km
    b) 5°C per km
    c) 6.5°C per km
    d) 10°C per km
28. Temperature inversion is a condition where
    a) Temperature increases with height
    b) Temperature remains constant
    c) Temperature decreases with height
    d) Humidity increases with height
29. Which layer is associated with temperature inversion during winter nights?
    a) Stratosphere
    b) Troposphere
    c) Mesosphere
    d) Exosphere

30. Which layer absorbs harmful ultraviolet radiation?
    a) Troposphe b) Stratosphere
    c) Mesosphere d) Thermosphere
31. Ozone in the stratosphere protects life by absorbing
    a) Infrared rays b) Ultraviolet rays
    c) Visible light d) Gamma rays
32. Weather processes critical to agriculture occur in
    a) Exosphere b) Stratosphere
    c) Troposphere d) Mesosphere
33. The greenhouse effect is mainly caused by
    a) Dust particles
    b) Ozone
    c) Greenhouse gases like CO2 and CH□
    d) Oxygen
34. Carbon dioxide is essential for plants because it is used in:
    a) Respiration b) Photosynthesis
    c) Transpiration d) Germination
35. Atmospheric pressure:
    a) Increases with altitude b) Decreases with altitude
    c) Remains constant d) Is highest at mountain peaks
36. The ionosphere is part of the:
    a) Troposphere b) Stratosphere
    c) Thermosphere d) Mesosphere
37. Temperature increases sharply in the thermosphere due to:
    a) Infrared radiation b) Absorption of UV & X-rays
    c) CO_2 presence d) Cloud formation
38. Which gas, though present in small amounts, is crucial for weather and climate?
    a) Nitrogen b) Oxygen
    c) Carbon dioxide d) Argon
39. Jet streams are found in the
    a) Mesosphere b) Lower stratosphere
    c) Thermosphere d) Exosphere

40. The exosphere mainly consists of
    a) Oxygen and nitrogen
    b) Hydrogen and helium
    c) Carbon dioxide
    d) Water vapor
41. The ozone hole is primarily caused by
    a) Carbon monoxide
    b) Chlorofluorocarbons (CFCs)
    c) Nitrogen oxides
    d) Carbon dioxide
42. Which human activity most contributes to greenhouse gases?
    a) Irrigation
    b) Combustion of fossil fuels
    c) Fertilizer application
    d) Ploughing
43. The atmospheric layer responsible for cloud formation is
    a) Troposphere
    b) Stratosphere
    c) Thermosphere
    d) Exosphere
44. Satellites for weather monitoring orbit in the
    a) Mesosphere
    b) Thermosphere
    c) Stratosphere
    d) Troposphere
45. The Earth's atmosphere was formed
    a) 500 million years ago
    b) 2 billion years ago
    c) 4.6 billion years ago
    d) 10 billion years ago
46. Ozone layer is available in
    a) Troposphere
    b) Stratosphere
    c) Mesosphere
    d) Thermosphere
47. Which of the following is true about the mesosphere?
    a) Hottest layer
    b) Coldest layer
    c) No change in temperature
    d) Contains ozone maximum
48. The density of air is maximum at
    a) Sea level
    b) 10 km height
    c) Tropopause
    d) Stratosphere
49. The Earth's atmosphere is densest in the
    a) Troposphere
    b) Mesosphere
    c) Thermosphere
    d) Exosphere
50. Which layer merges into outer space?
    a) Stratosphere
    b) Mesosphere
    c) Exosphere
    d) Thermosphere

Answer Key

1	c	2	b	3	c	4	c	5	b	6	a	7	b
8	c	9	c	10	d	11	d	12	d	13	b	14	c
15	c	16	b	17	b	18	b	19	b	20	c	21	a
22	c	23	c	24	a	25	c	26	a	27	c	28	a
29	b	30	b	31	b	32	c	33	c	34	b	35	b
36	c	37	b	38	c	39	b	40	b	41	b	42	b
43	a	44	b	45	c	46	b	47	b	48	a	49	a
50	c												

Ch. 3. Atmospheric Pressure and Wind

1. Atmospheric pressure at sea level is approximately
   a) 760 mm Hg b) 1013.25 hPa
   c) 1 atm d) All of the above
2. With increasing altitude, atmospheric pressure
   a) Increases b) Decreases
   c) Remains constant d) First increases then decreases
3. The instrument used to measure atmospheric pressure is
   a) Anemometer b) Hygrometer
   c) Barometer d) Thermometer
4. One hPa (hectopascal) is equal to:
   a) 1 N/m^2 b) 1000 Pa
   c) 100 Pa d) 10 Pa
5. Standard atmospheric pressure in mb (millibar) is
   a) 1000 b) 760
   c) 1013.25 d) 900
6. Pressure decreases approximately by 1 hPa for every
   a) 5 m increase in height b) 8.5 m
   c) 10 m d) 12 m
7. The vertical decrease of pressure with height is called
   a) Lapse rate b) Pressure gradient
   c) Hydrostatic gradient d) Barometric gradient
8. High pressure is associated with:
   a) Rising air b) Clear skies
   c) Cyclones d) Cloud formation

9. Pressure is lowest at:
   a) Sea level b) Mount Everest
   c) Dead Sea d) Tropopause
10. Isobars are lines joining points of equal:
   a) Temperature b) Humidity
   c) Pressure d) Wind speed
11. Wind is caused primarily by:
   a) Earth's rotation
   b) Unequal heating of Earth's surface
   c) Ocean currents
   d) Earth's revolution
12. Anemometer measures:
   a) Wind speed b) Wind direction
   c) Air temperature d) Humidity
13. Wind vane indicates:
   a) Wind speed b) Wind direction
   c) Wind pressure d) None
14. Wind moving from high to low pressure and deflected due to Earth's rotation is affected by:
   a) Frictional force b) Coriolis force
   c) Centrifugal force d) Gravitational force
15. Coriolis effect causes wind to deflect:
   a) Right in Northern Hemisphere b) Left in Southern Hemisphere
   c) Opposite in each hemisphere d) Both a and b
16. The general wind circulation between 30° and 60° latitudes is known as:
   a) Trade winds b) Westerlies
   c) Polar easterlies d) Monsoon winds
17. Trade winds blow from:
   a) Subtropical high to equatorial low
   b) Equator to poles
   c) Poles to equator
   d) High latitudes to low latitudes

18. Local winds include all EXCEPT
    a) Loo
    b) Chinook
    c) Jet stream
    d) Sea breeze
19. Doldrums refer to
    a) Area of calm near the equator
    b) Cold polar winds
    c) Deserts
    d) Monsoon zone
20. Jet streams are
    a) Fast upper-level winds
    b) Surface winds
    c) Seasonal winds
    d) Local winds
21. Sea breeze occurs during
    a) Night
    b) Daytime
    c) Winter
    d) Morning only
22. Land breeze occurs when
    a) Sea is cooler at night
    b) Land is warmer at night
    c) Land cools faster than sea at night
    d) Sea heats faster than land at night
23. Wind speed is generally higher during
    a) Night
    b) Early morning
    c) Afternoon
    d) Dusk
24. Monsoons are caused by
    a) Ocean currents
    b) Seasonal pressure differences
    c) Earth's rotation
    d) Frontal lifting
25. Winter monsoon winds in India flow from
    a) Ocean to land
    b) Land to ocean
    c) North to south
    d) East to west
26. Wind speed is affected by
    a) Pressure gradient
    b) Friction
    c) Coriolis force
    d) All of the above
27. Diurnal wind variation is most noticeable near
    a) Mountains
    b) Oceans
    c) Coastal areas
    d) Deserts

28. Katabatic winds flow
    a) Downhill at night   b) Uphill during day
    c) Across oceans   d) None of the above
29. Seasonal wind reversal is typical of
    a) Equatorial regions   b) Tropical deserts
    c) Monsoon climates   d) Polar regions
30. Wind chill effect is more noticeable at
    a) Higher humidity   b) Higher wind speeds
    c) Higher temperatures   d) Lower altitudes
31. Cyclones are associated with
    a) High pressure   b) Sinking air
    c) Low pressure and rising air   d) Clear skies
32. Cyclonic circulation is
    a) Clockwise in Northern Hemisphere
    b) Counter-clockwise in Northern Hemisphere
    c) Same in both hemispheres
    d) Anticlockwise in Southern Hemisphere
33. Anticyclones are associated with
    a) Cloud formation   b) Low pressure
    c) High pressure and clear skies   d) Storms
34. Cyclones in the Indian Ocean are called
    a) Hurricanes   b) Typhoons
    c) Tropical cyclones   d) Tornadoes
35. Hurricanes form over
    a) Cold oceans   b) Warm tropical oceans
    c) Mountains   d) Deserts
36. Tropical cyclones are fueled by
    a) Cold air aloft   b) Condensation of water vapor
    c) Jet streams   d) Cold ocean currents
37. An anticyclone causes
    a) Heavy rainfall   b) Thunderstorms
    c) Dry and fair weather   d) Cyclogenesis

38. Eye of a cyclone is
    a) Area of strongest winds
    b) Calm and low-pressure area
    c) High-pressure zone
    d) Always cloudy
39. Cyclones rotate counter-clockwise in:
    a) Southern Hemisphere
    b) Northern Hemisphere
    c) Both hemispheres
    d) At poles
40. Pressure at the center of an anticyclone is:
    a) Lowest
    b) Highest
    c) Same as surrounding
    d) Undefined
41. Sea breeze generally begins:
    a) Just before sunrise
    b) Late morning
    c) Noon or afternoon
    d) Midnight
42. Land breeze blows from:
    a) Ocean to land
    b) Land to ocean
    c) Mountain to valley
    d) South to north
43. Sea breeze strength depends on:
    a) Wind speed
    b) Temperature difference
    c) Pressure difference
    d) Both B and C
44. Local winds can affect:
    a) Crop transpiration
    b) Pollination
    c) Evaporation rate
    d) All of the above
45. The reversal of land and sea breeze is due to
    a) Coriolis force
    b) Difference in heating rate of land and water
    c) Gravity
    d) Wind direction
46. Which wind system influences Indian agriculture the most?
    a) Westerlies
    b) Monsoon
    c) Easterlies
    d) Trade winds
47. Anticyclones are beneficial for agriculture when
    a) Crops need sunshine
    b) Humidity is high
    c) Rain is needed
    d) During monsoon

48. High wind speeds can damage
    a) Root crops b) Tall crops
    c) Legumes d) All types equally
49. Wind erosion is most common in
    a) Wetlands b) Deserts
    c) Mountains d) Tropical forests
50. Windbreaks in agriculture help in
    a) Increasing evaporation b) Reducing wind erosion
    c) Decreasing yield d) Attracting cyclones

Answer Key

1	d	2	b	3	c	4	c	5	c	6	c	7	c
8	d	9	b	10	c	11	b	12	a	13	b	14	b
15	d	16	b	17	a	18	c	19	a	20	a	21	b
22	c	23	c	24	b	25	b	26	d	27	c	28	a
29	c	30	b	31	c	32	b	33	c	34	c	35	b
36	b	37	c	38	b	39	b	40	b	41	c	42	b
43	d	44	d	45	b	46	b	47	a	48	b	49	b
50	b												

Ch. 4. Solar Radiation

1. The primary source of energy for Earth's atmosphere is:
    a) Earth's core b) Moon
    c) Sun d) Volcanoes
2. Solar radiation is also known as:
    a) Geothermal radiation b) Insolation
    c) Radiation balance d) Ozone
3. Radiation emitted by the sun is mainly:
    a) Longwave b) Ultraviolet only
    c) Shortwave d) Infrared
4. The solar radiation reaching the top of Earth's atmosphere is known as:
    a) Terrestrial radiation b) Solar constant
    c) Albedo d) Scattered radiation
5. Solar energy is transferred in the form of:
    a) Convection b) Conduction
    c) Radiation d) Advection

6. The accepted value of the solar constant is approximately:
   a) 100 W/m²   b) 340 W/m²
   c) 1353 W/m²   d) 2000 W/m²
7. The unit of solar radiation is:
   a) Lux
   b) Watts per square meter (W/m²)
   c) Kelvin
   d) Joules per mole
8. The solar constant is measured at a distance of:
   a) 100 km from Earth   b) Sea level
   c) Outside Earth's atmosphere   d) On Earth's surface
9. Variation in the solar constant is mainly due to:
   a) Earth's magnetic field   b) Atmospheric pressure
   c) Sunspot activity   d) Greenhouse gases
10. Solar constant represents:
    a) Net radiation at Earth's surface
    b) Maximum solar radiation on cloudy day
    c) Average radiation at top of atmosphere
    d) Amount of heat in Earth's core
11. The reduction of solar radiation before reaching Earth's surface is known as:
    a) Reflection   b) Depletion
    c) Refraction   d) Transmission
12. The major causes of solar radiation depletion include:
    a) Absorption   b) Scattering
    c) Reflection   d) All of the above
13. Most solar radiation is absorbed by:
    a) Nitrogen   b) Water vapor and ozone
    c) Oxygen   d) CO_2
14. Scattering of solar radiation causes
    a) Rainbow   b) Blue sky
    c) Heating of surface   d) Dark nights

15. Ozone in the stratosphere mainly absorbs
    a) Infrared radiation
    b) Ultraviolet radiation
    c) Microwave radiation
    d) Visible light
16. Radiation from the sun is mostly:
    a) Longwave
    b) Microwave
    c) Shortwave
    d) Thermal infrared
17. Radiation emitted by Earth's surface is known as
    a) Solar radiation
    b) Terrestrial or longwave radiation
    c) Direct radiation
    d) Reflected radiation
18. Longwave radiation refers to
    a) 0.1 – 0.4 μm
    b) 0.4 – 0.7 μm
    c) Above 4 μm
    d) Below 1 μm
19. Shortwave radiation falls in the range of
    a) 1 – 10 μm
    b) 0.3 – 3 μm
    c) 10 – 100 μm
    d) 5 – 50 μm
20. Thermal radiation is associated with
    a) Photosynthesis
    b) Longwave radiation
    c) UV radiation
    d) Ozone depletion
21. Net radiation is the difference between
    a) Day and night temperature
    b) Incoming and outgoing radiation
    c) Solar and lunar radiation
    d) Shortwave and longwave radiation
22. During daytime, net radiation is usually
    a) Zero
    b) Negative
    c) Positive
    d) Constant
23. At night, net radiation is typically
    a) Positive
    b) Zero
    c) Negative
    d) Infinite
24. Which of the following contributes to outgoing radiation?
    a) Reflected solar radiation
    b) Emitted longwave radiation
    c) Both a and b
    d) Only direct radiation

25. Net radiation drives processes like
    a) Evaporation b) Convection
    c) Crop growth d) All of the above
26. Albedo is defined as
    a) Amount of radiation absorbed
    b) Amount of radiation emitted
    c) Ratio of reflected to incident solar radiation
    d) Ratio of longwave to shortwave radiation
27. Fresh snow has an albedo of approximately:
    a) 5% b) 20%
    c) 40% d) 80–90%
28. Dark-colored surfaces have:
    a) Higher albedo b) Lower albedo
    c) Constant albedo d) No albedo
29. Vegetated surfaces generally reflect about:
    a) 10–20% b) 40–50%
    c) 60–70% d) 80%
30. Which of the following has the highest albedo?
    a) Grassland b) Forest
    c) Snow d) Water
31. Solar radiation is essential for:
    a) Crop respiration b) Photosynthesis
    c) Nutrient uptake d) Soil erosion
32. The most important band of radiation for photosynthesis lies in:
    a) Ultraviolet b) Infrared
    c) Visible light d) X-rays
33. The peak wavelength of solar radiation is near:
    a) 1 μm b) 0.5 μm
    c) 0.3 μm d) 2.0 μm
34. PAR (Photosynthetically Active Radiation) lies in the range of
    a) 400–700 nm b) 100–300 nm
    c) 700–1000 nm d) 10–20 μm

35. Solar radiation affects all except
    a) Crop maturity
    b) Plant transpiration
    c) Leaf pigment
    d) Soil porosity
36. Cloudy skies generally result in
    a) More incoming shortwave radiation
    b) Less incoming radiation
    c) Higher surface temperature
    d) No change in radiation
37. Which surface reflects the least amount of radiation?
    a) Snow
    b) Dry soil
    c) Black soil
    d) Water
38. Clear nights are colder due to
    a) Radiation gain
    b) Radiation fog
    c) Loss of longwave radiation
    d) Increased albedo
39. Day length affects the amount of
    a) Net radiation
    b) Albedo
    c) Longwave radiation
    d) Ozone
40. Which factor does NOT affect solar radiation at surface level?
    a) Latitude
    b) Altitude
    c) Soil moisture
    d) Cloud cover
41. Radiation balance is a key factor in
    a) Rainfall prediction
    b) Surface temperature regulation
    c) Soil profile development
    d) Wind pattern creation
42. Radiation inversion occurs
    a) During midday
    b) On cloudy nights
    c) On clear, calm nights
    d) During rainfall
43. Scattering is maximum when particles are
    a) Larger than wavelength
    b) Smaller than wavelength
    c) Same size as wavelength
    d) Heavier
44. The green-house effect is primarily related to
    a) Shortwave radiation
    b) Ultraviolet radiation
    c) Longwave radiation
    d) X-rays

45. Reflected radiation increases with:
    a) Cloudiness b) Surface roughness
    c) Surface brightness d) Soil fertility
46. Which of these gases is transparent to solar radiation but opaque to terrestrial radiation?
    a) Oxygen b) Nitrogen
    c) Carbon dioxide d) Argon
47. Net radiation is negative when:
    a) Incoming = outgoing radiation
    b) Outgoing radiation exceeds incoming
    c) No clouds are present
    d) Albedo is high
48. The Stefan-Boltzmann law relates radiation to:
    a) Pressure b) Humidity
    c) Temperature d) Wind speed
49. Which radiation law gives the wavelength of maximum emission?
    a) Planck's law b) Stefan-Boltzmann law
    c) Wien's displacement law d) Newton's law
50. Which surface emits more longwave radiation at night?
    a) Grass b) Concrete
    c) Wet soil d) All emit equally

Answer Key

1	c	2	b	3	c	4	b	5	c	6	c	7	b
8	c	9	c	10	c	11	b	12	d	13	b	14	b
15	b	16	c	17	b	18	c	19	b	20	b	21	b
22	c	23	c	24	c	25	d	26	c	27	d	28	b
29	a	30	c	31	b	32	c	33	b	34	a	35	d
36	b	37	c	38	c	39	a	40	c	41	b	42	c
43	b	44	c	45	c	46	c	47	b	48	c	49	c
50	a												

Ch. 5. Light - A Comprehensive Overview

1. Light is a form of
   a) Sound energy
   b) Mechanical energy
   c) Electromagnetic energy
   d) Chemical energy
2. The speed of light in vacuum is:
   a) 3×10^8 m/s
   b) 3×10^8 m/s
   c) 3×10^8 m/s
   d) 3×10^8 m/s
3. Light waves are:
   a) Longitudinal
   b) Transverse
   c) Mechanical
   d) Sound
4. The dual nature of light means it behaves like:
   a) Wave only
   b) Particle only
   c) Both wave and particle
   d) None
5. Which of the following proves the wave nature of light?
   a) Photoelectric effect
   b) Interference
   c) Compton effect
   d) Reflection
6. The law of reflection states:
   a) Angle of incidence = angle of reflection
   b) Angle of incidence = angle of refraction
   c) Both angles are 90°
   d) None
7. The image formed by a plane mirror is:
   a) Real and inverted
   b) Virtual and inverted
   c) Real and erect
   d) Virtual and erect
8. In concave mirrors, when the object is at the center of curvature, the image is:
   a) Real, inverted, same size
   b) Virtual, erect, smaller
   c) Virtual, erect, larger
   d) Real, inverted, diminished
9. Rearview mirrors in vehicles use:
   a) Plane mirror
   b) Concave mirror
   c) Convex mirror
   d) None
10. The focal length of a plane mirror is:
    a) 0
    b) Infinity
    c) Positive
    d) Negative

11. Refraction occurs due to:
    a) Change in speed of light b) Change in frequency
    c) Reflection d) Polarization
12. The bending of light when it enters a denser medium is:
    a) Away from the normal b) Toward the normal
    c) Parallel to the surface d) None
13. The lens that converges parallel rays is:
    a) Concave b) Convex
    c) Cylindrical d) None
14. Focal length of a convex lens is:
    a) Positive b) Negative
    c) Zero d) Variable
15. Unit of power of a lens:
    a) Watt b) Diopter
    c) Meter d) Hertz
16. Formula for lens power (P):
    a) $P = 1/f$ b) $P = f$
    c) $P = f^2$ d) $P = 1/f^2$
17. Refractive index (n) is given by:
    a) Speed in vacuum/speed in medium
    b) Speed in medium/speed in vacuum
    c) Distance/time
    d) None
18. Critical angle occurs when:
    a) Angle of refraction is 0° b) Angle of incidence is 90°
    c) Angle of refraction is 90° d) Total internal reflection starts
19. Which lens is used to correct hypermetropia?
    a) Concave b) Convex
    c) Plane d) Cylindrical
20. Which lens is used to correct myopia?
    a) Concave b) Convex
    c) Plane d) None

21. Splitting of light into colors is known as
    a) Refraction b) Diffraction
    c) Dispersion d) Scattering
22. In dispersion of white light, the color that bends the most is:
    a) Red b) Violet
    c) Green d) Blue
23. Rainbow formation is due to
    a) Reflection only
    b) Refraction only
    c) Total internal reflection
    d) Refraction, reflection and dispersion
24. Young's double-slit experiment shows
    a) Particle nature b) Quantum nature
    c) Wave nature d) Nuclear nature
25. Diffraction is the:
    a) Spreading of waves at obstacle edges
    b) Refraction of light
    c) Dispersion of rays
    d) Polarization of light
26. Photosynthetically Active Radiation (PAR) lies in range
    a) 100–400 nm b) 400–700 nm
    c) 700–900 nm d) 200–400 nm
27. Light affects which of the following plant activities most?
    a) Germination b) Transpiration
    c) Photosynthesis d) Respiration
28. Which type of light is most effective for photosynthesis?
    a) Green b) Blue and red
    c) Yellow d) Infrared
29. In agriculture, photoperiodism is associated with
    a) Wind speed b) Light duration
    c) Soil pH d) Humidity
30. Light intensity in agriculture is measured using
    a) Lux meter b) Barometer
    c) Anemometer d) Thermometer

31. Instrument to view distant objects clearly
    a) Microscope b) Telescope
    c) Periscope d) Spectrometer
32. Magnification of simple microscope is:
    a) D/f b) f/D
    c) f + D d) D × f
33. Which part of the eye controls the amount of light entering?
    a) Retina b) Lens
    c) Iris d) Cornea
34. A person who cannot see distant objects suffers from:
    a) Hypermetropia b) Myopia
    c) Astigmatism d) Cataract
35. Which optical defect is due to aging?
    a) Hypermetropia b) Myopia
    c) Presbyopia d) Astigmatism
36. Light year is a unit of:
    a) Time b) Distance
    c) Speed d) Mass
37. Convex lens used in a projector forms image that is:
    a) Real and inverted b) Virtual and erect
    c) Real and erect d) Virtual and inverted
38. For total internal reflection, light must travel from:
    a) Denser to rarer medium b) Rarer to denser
    c) Vacuum to air d) Water to air only
39. The angle between reflected & incident ray is 60°. Angle of incidence is:
    a) 60° b) 30°
    c) 45° d) 90°
40. Luminous intensity is measured in:
    a) Candela b) Lux
    c) Lumen d) Watt
41. The visible spectrum consists of:
    a) 100–400 nm b) 400–700 nm
    c) 700–1000 nm d) 200–600 nm

42. Color of an object is due to:
    a) Light absorbed b) Light reflected
    c) Light transmitted d) Refracted light
43. Light from the sun reaches Earth in approx:
    a) 8 minutes b) 8 seconds
    c) 1 minute d) 24 minutes
44. In a lens, the point where parallel rays converge is called:
    a) Center b) Pole
    c) Focal point d) Principal axis
45. Real image can be captured on a:
    a) Retina b) Paper/screen
    c) Mirror d) Convex lens only
46. Light rays do not bend when entering a medium at:
    a) 90° b) 60°
    c) 45° d) 30°
47. The reason why objects appear inverted in a spoon is:
    a) Concave reflection b) Plane reflection
    c) Dispersion d) Diffraction
48. Light behaves as a particle in:
    a) Reflection b) Refraction
    c) Photoelectric effect d) Interference
49. The sunlight splits into seven colors due to:
    a) Refraction b) Dispersion
    c) Reflection d) Polarization
50. White light is composed of how many colors?
    a) 3 b) 5
    c) 7 d) 9

Answer Key

1	c	2	c	3	b	4	c	5	b	6	a	7	d
8	a	9	c	10	a	11	a	12	b	13	b	14	a
15	b	16	a	17	a	18	c	19	b	20	a	21	c
22	b	23	d	24	c	25	a	26	b	27	c	28	b
29	b	30	a	31	b	32	a	33	c	34	b	35	c
36	b	37	a	38	a	39	b	40	a	41	b	42	b
43	a	44	c	45	b	46	a	47	a	48	c	49	b
50	c												

Ch. 6. Atmospheric Temperature

1. Temperature is a measure of
   a) Humidity in the air
   b) Heat content of a substance
   c) Average kinetic energy of molecules
   d) Potential energy of molecules
2. The instrument used to measure atmospheric temperature is:
   a) Anemometer b) Thermometer
   c) Barometer d) Hygrometer
3. Maximum temperature of the day is usually recorded at:
   a) Sunrise b) 12 noon
   c) 2–3 PM d) Sunset
4. The minimum temperature of the day generally occurs:
   a) At midnight b) Just before sunrise
   c) At noon d) In the evening
5. Standard screen height for measuring air temperature is:
   a) 1 m b) 1.5 m
   c) 2.0 m d) 0.5 m
6. Lapse rate is the rate at which:
   a) Wind speed decreases with altitude
   b) Pressure increases with height
   c) Temperature decreases with height
   d) Rainfall decreases with time

7. The normal lapse rate in the troposphere is approximately:
   a) 3.5°C/km b) 6.5°C/km
   c) 9.8°C/km d) 12°C/km
8. The dry adiabatic lapse rate is
   a) 6.5°C/km b) 9.8°C/km
   c) 4.5°C/km d) 10.5°C/km
9. Moist adiabatic lapse rate is generally lower because of
   a) Heat absorption by ozone
   b) Latent heat release during condensation
   c) Increased pressure with height
   d) Decreased CO_2 concentration
10. Lapse rate is positive when temperature
   a) Increases with altitude b) Decreases with altitude
   c) Remains constant d) Is at maximum
11. Temperature inversion refers to a situation where
   a) Temperature increases with altitude
   b) Temperature remains constant
   c) Temperature decreases rapidly
   d) Rainfall occurs
12. Which of the following favors temperature inversion?
   a) Windy night b) Cloudy sky
   c) Clear and calm night d) Summer afternoon
13. Radiation inversion typically occurs during
   a) Daytime b) Cloudy days
   c) Clear, calm nights d) Rainy days
14. Temperature inversion is important in agriculture because it can lead to:
   a) Drought b) Enhanced crop growth
   c) Frost damage d) Wind erosion
15. Subsidence inversion is commonly associated with
   a) Tropical storms b) High-pressure systems
   c) Cyclonic storms d) Thunderstorms

16. Daily range of temperature is the difference between
    a) Summer and winter temperature
    b) Maximum and minimum temperature in a day
    c) Latitude and longitude temperature
    d) Urban and rural temperature
17. In deserts, the daily range of temperature is
    a) Low b) Negligible
    c) High d) Constant
18. Seasonal temperature variation is mainly due to
    a) Moon phases b) Earth's rotation
    c) Tilt of Earth's axis & revolution d) Volcanic activity
19. The hottest month in Northern India is usually
    a) April b) May
    c) June d) July
20. In India, minimum temperatures are lowest in
    a) March b) December
    c) January d) February
21. In the troposphere, temperature
    a) Increases with height b) Decreases with height
    c) Remains constant d) Drops sharply only at 10 km
22. Temperature increases with altitude in
    a) Troposphere b) Stratosphere
    c) Mesosphere d) All layers
23. The layer of atmosphere with coldest temperatures is
    a) Troposphere b) Mesosphere
    c) Stratosphere d) Thermosphere
24. Which atmospheric layer contains the ozone layer, affecting temperature structure?
    a) Mesosphere b) Stratosphere
    c) Thermosphere d) Exosphere
25. The temperature in the thermosphere
    a) Drops continuously b) Increases rapidly
    c) Remains constant d) Is very low

26. Latitude affects temperature because it determines
    a) Cloud cover b) Wind speed
    c) Sun's angle of incidence d) Ocean currents
27. Proximity to water bodies causes
    a) Increase in temperature b) High diurnal range
    c) Lower temperature range d) Temperature inversion
28. Urban areas are warmer than rural areas due to
    a) Lower albedo b) Heat absorption by concrete
    c) Less vegetation d) All of the above
29. Altitude affects temperature such that
    a) Higher areas are warmer b) Lower areas are cooler
    c) Temperature decreases with height d) Height has no effect
30. Which surface heats and cools faster?
    a) Water b) Sand
    c) Vegetation d) Forest
31. Temperature influences crop growth by affecting:
    a) Germination b) Flowering
    c) Maturity d) All of the above
32. Optimum temperature for most crops lies between:
    a) 10–15°C b) 15–25°C
    c) 25–35°C d) 35–45°C
33. Thermal time is calculated using
    a) Humidity and rainfall b) Wind and sunshine
    c) Temperature and crop threshold d) Evaporation & transpiration
34. Growing Degree Days (GDD) concept is based on:
    a) Wind speed b) Solar radiation
    c) Daily temperature d) Humidity index
35. Frost injury to crops occurs during:
    a) Daytime b) High humidity
    c) Temperature inversion d) Cloudy nights
36. Zero degree Celsius is equal to:
    a) 273 K b) 0 K
    c) 32 K d) 100 K

37. The SI unit of temperature is:
    a) Celsius b) Kelvin
    c) Fahrenheit d) Degree
38. Which gas absorbs terrestrial radiation and helps in warming lower atmosphere?
    a) Oxygen b) Nitrogen
    c) Carbon dioxide d) Argon
39. Which temperature scale has no negative values?
    a) Celsius b) Kelvin
    c) Fahrenheit d) Reaumur
40. Inversion layer prevents
    a) Cloud formation b) Air mixing
    c) Fog formation d) Radiation loss
41. Temperature variation near soil surface is more during
    a) Winter b) Summer
    c) Cloudy season d) Rainy season
42. High-altitude stations are cooler because of
    a) Strong winds b) Less water vapor
    c) Adiabatic cooling d) Cloud cover
43. Which factor does NOT influence surface air temperature?
    a) Solar radiation b) Wind direction
    c) Soil pH d) Cloud cover
44. Night temperature drops quickly under
    a) Windy and cloudy conditions b) Clear and calm conditions
    c) Humid atmosphere d) None of the above
45. Thermal belt in hilly regions is a zone of
    a) Maximum humidity b) High rainfall
    c) Least frost occurrence d) Maximum inversion
46. An adiabatic process occurs without
    a) Change in volume b) Exchange of heat
    c) Change in temperature d) Air movement
47. Cold air drains to valley bottoms at night, causing
    a) Rain b) Fog
    c) Frost pockets d) Warm air uplift

48. Vertical temperature gradient in stratosphere is
    a) Positive b) Negative
    c) Neutral d) Irregular
49. Thermal inversion may lead to:
    a) Cloud formation b) Smog accumulation
    c) Increased rainfall d) Thunderstorm
50. A sudden drop in temperature may be harmful during:
    a) Harvest b) Pollination
    c) Germination d) Irrigation

Answer Key

1	c	2	c	3	b	4	c	5	b	6	a	7	d
8	a	9	c	10	a	11	a	12	b	13	b	14	a
15	b	16	a	17	a	18	c	19	b	20	c	21	b
22	b	23	d	24	c	25	a	26	b	27	c	28	b
29	b	30	a	31	b	32	a	33	c	34	b	35	c
36	b	37	a	38	a	39	b	40	a	41	b	42	b
43	a	44	c	45	b	46	a	47	a	48	c	49	b
50	b												

Ch.7. Thermal Regime and Crop Growth & Development

1. Thermal time is often expressed in terms of
    a) Relative humidity
    b) Growing Degree Days (GDD)
    c) Precipitation index
    d) Wind speed
2. Growing Degree Days (GDD) are calculated based on:
    a) Soil temperature b) Daily rainfall
    c) Daily mean air temperature d) Daily maximum humidity
3. Base temperature is the:
    a) Maximum crop temperature
    b) Temperature below which growth ceases
    c) Mean annual temperature
    d) Soil surface temperature

4. GDD = (Tmax + Tmin)/2 - Tb. Here, Tb is:
   a) Top soil temperature b) Base temperature
   c) Bulk temperature d) Bright sunshine
5. For wheat, the base temperature commonly used is:
   a) 0°C b) 5°C
   c) 10°C d) 15°C
6. Thermal time requirement is expressed in units of:
   a) Degree Celsius b) Hours
   c) Degree Days d) Kelvin
7. GDD is used to predict:
   a) Rainfall b) Wind speed
   c) Crop phenology d) Evaporation rate
8. Which of the following crops has the lowest base temperature?
   a) Maize b) Rice
   c) Potato d) Wheat
9. Thermal time accumulation is useful in:
   a) Sowing date prediction b) Fertilizer scheduling
   c) Pest emergence forecasting d) Both A and C_2
10. Which is not a factor affecting GDD accuracy?
   a) Base temperature selection b) Rainfall pattern
   c) Accurate temperature recording d) Crop variety
11. Crop heat units help in assessing
   a) Soil salinity b) Crop water needs
   c) Crop growth stages d) Market price
12. Which stage of crop growth is most sensitive to thermal time?
   a) Germination b) Vegetative
   c) Reproductive d) All of the above
13. Thermal time is useful in areas with
   a) Constant rainfall b) Constant sunshine
   c) Variable temperatures d) Uniform topography
14. Higher temperatures than optimal base temperature lead to
   a) Faster development b) Better yield
   c) Frost injury d) Slower growth

15. If the average temperature is less than base temperature, GDD is
    a) Zero
    b) Negative
    c) Equal to base temperature
    d) Positive
16. Pest emergence models are often based on
    a) Rainfall and humidity
    b) GDD and thresholds
    c) Solar radiation
    d) Wind speed
17. Which insect is commonly modeled using heat units?
    a) Grasshopper
    b) Aphid
    c) Stem borer
    d) All of the above
18. Thermal time for pests is calculated similar to
    a) Temperature lapse rate
    b) Solar intensity
    c) Crop GDD
    d) Wind direction
19. Insects develop faster at
    a) High rainfall
    b) Below base temperatures
    c) Above base temperatures
    d) Freezing temperatures
20. Integrated pest forecasting tools combine
    a) Weather and soil type
    b) Thermal time and pest thresholds
    c) Water balance and humidity
    d) Fertilizer and rainfall data
21. Crop weather calendars indicate
    a) Daily temperature
    b) Expected crop stages vs time
    c) Fertilizer needs
    d) Pesticide brands
22. A crop calendar is helpful for
    a) Harvest planning
    b) Identifying weather risks
    c) Advisory services
    d) All of the above
23. Pest weather calendars are useful to
    a) Estimate pest population
    b) Plan pesticide spray
    c) Issue early warnings
    d) All of the above
24. Crop-weather calendars are usually developed at
    a) International level
    b) Zonal level
    c) Farm level
    d) Agro-climatic zone level

25. In crop calendars, phenological stages are matched with
    a) Wind speed
    b) Solar radiation
    c) Thermal time and weather conditions
    d) Market demand
26. The main source of energy for Earth is
    a) Geothermal heat   b) The moon
    c) The sun   d) Fossil fuels
27. The solar constant is approximately
    a) 1270 W/m^2 b) 1361 W/m^2
    c) 1013 W/m^2 d) 980 W/m^2
28. Albedo is a measure of:
    a) Absorption   b) Reflection
    c) Transmission   d) Emission
29. Snow has a high albedo of around
    a) 0.1   b) 0.4
    c) 0.8   d) 1.0
30. Earth's surface emits
    a) Shortwave radiation   b) Longwave radiation
    c) Microwave radiation   d) Gamma rays
31. Incoming solar radiation is known as:
    a) Terrestrial radiation   b) Net radiation
    c) Insolation   d) Emitted energy
32. Outgoing radiation is largely in the form of
    a) UV   b) Infrared
    c) X-rays   d) Visible light
33. Net radiation is the difference between
    a) Radiation and convection
    b) Incoming and outgoing radiation
    c) Reflected and absorbed radiation
    d) Latent and sensible heat
34. Which component does not affect Earth's radiation balance?
    a) Albedo   b) Cloud cover
    c) Latitude   d) Soil type

35. The greenhouse effect is related to:
    a) Longwave radiation trapping   b) UV radiation
    c) Ocean current circulation   d) Soil moisture loss
36. A crop with high GDD requirement is likely to:
    a) Mature early   b) Require cool climate
    c) Have a longer growing season   d) Grow in desert areas only
37. Which weather parameter is not required for GDD calculation?
    a) Maximum temperature   b) Minimum temperature
    c) Base temperature   d) Rainfall
38. Which tool can assist in GDD computation?
    a) Rain gauge   b) Thermometer
    c) Crop simulation models   d) Soil auger
39. The concept of GDD fails when
    a) Temperature exceeds threshold
    b) Crop is irrigated
    c) Base temperature is wrongly selected
    d) Crop is fertilized
40. Which layer balances most of the Earth's energy inputs and outputs?
    a) Stratosphere   b) Troposphere
    c) Lithosphere   d) Exosphere
41. Thermal time accumulation stops during
    a) Photosynthesis   b) Rainy season
    c) Dormancy   d) Flowering
42. Evapotranspiration demand is linked to:
    a) Thermal time   b) Soil color
    c) GDD   d) Both a and c
43. What is the unit of solar constant?
    a) Cal/m^2 b) W/m^2
    c) Lux   d) Degree days
44. Energy loss from Earth to atmosphere is mainly by:
    a) Reflection   b) Longwave radiation
    c) UV scattering   d) Ground heat flux

45. High albedo means:
    a) High absorption
    b) High reflection
    c) High temperature
    d) Low reflection
46. Farmers can use crop weather calendars for:
    a) Irrigation scheduling
    b) Planting decisions
    c) Pest management
    d) All of the above
47. Thermal time ensures which aspect of plant growth?
    a) Predictable stage duration
    b) Marketability
    c) Disease resistance
    d) Genetic variability
48. Solar energy reaching Earth's surface is about:
    a) 100%
    b) 70%
    c) 50%
    d) 30%
49. Clouds affect energy balance by:
    a) Absorbing radiation
    b) Reflecting sunlight
    c) Blocking infrared
    d) All of the above
50. The pest-weather calendar is a tool in:
    a) Precision irrigation
    b) Agro-meteorological advisory services
    c) Soil fertility estimation
    d) Seed germination testing

Answer Key

1	b	2	c	3	b	4	b	5	b	6	c	7	c
8	d	9	d	10	b	11	c	12	d	13	c	14	a
15	a	16	b	17	d	18	c	19	c	20	b	21	b
22	d	23	d	24	d	25	c	26	c	27	b	28	b
29	c	30	b	31	c	32	b	33	b	34	d	35	a
36	c	37	d	38	c	39	c	40	b	41	c	42	d
43	b	44	b	45	b	46	d	47	a	48	c	49	d
50	b												

Ch.8. Water Vapor in the Atmosphere

1. Humidity refers to
   a) Wind pressure b) Rainfall
   c) Water vapor in the air d) Temperature of the air
2. Relative humidity is expressed in
   a) Grams b) Percentage
   c) mm d) Watts
3. Relative Humidity (RH) is the ratio of
   a) Actual vapor pressure to saturation vapor pressure
   b) Dew point to temperature
   c) Wet bulb to dry bulb temperature
   d) Wind speed to air temperature
4. Absolute humidity is the
   a) Ratio of wet and dry bulb temperatures
   b) Mass of water vapor per unit volume of air
   c) Percentage of saturated air
   d) Temperature difference between dew and air
5. Saturation occurs when
   a) Air temperature rises
   b) Vapor pressure = Saturation vapor pressure
   c) Wind speed is zero
   d) Rain begins
6. Dew point is the temperature at which
   a) Wind stops b) Cloud forms
   c) Air becomes saturated d) Fog clears
7. When air becomes saturated, its RH is
   a) 50% b) 75%
   c) 90% d) 100%
8. Which instrument is used to measure relative humidity?
   a) Thermometer b) Hygrometer
   c) Rain gauge d) Barometer
9. RH is highest during
   a) Midday b) Early morning
   c) Afternoon d) Sunset

10. RH is lowest during
   a) Sunrise b) Noon
   c) Midnight d) Dawn
11. Vapour pressure is the pressure exerted by
   a) Air molecules
   b) Water molecules in vapor form
   c) Liquid water
   d) Solar radiation
12. Saturation vapour pressure increases with
   a) Decreasing temperature b) Increasing altitude
   c) Increasing temperature d) Decreasing pressure
13. Condensation is the process of
   a) Ice to water b) Water vapor to liquid water
   c) Liquid to gas d) Snow melting
14. Condensation releases
   a) Latent heat b) Kinetic energy
   c) Solar energy d) Wind force
15. Which is necessary for condensation to occur?
   a) High temperature b) Dry air
   c) Cooling of air to dew point d) Wind speed
16. Which particle helps in condensation?
   a) Oxygen b) Condensation nuclei
   c) Nitrogen d) Carbon dioxide
17. Condensation forms:
   a) Clouds b) Dew
   c) Fog d) All of the above
18. Which type of condensation forms on the ground surface?
   a) Clouds b) Dew
   c) Fog d) Haze
19. When condensation occurs close to the surface, it may lead to
   a) Cloud b) Snow
   c) Fog d) Hail

20. Latent heat released during condensation causes
    a) Cooling of air
    b) Warming of air
    c) Precipitation
    d) Cloud clearance
21. Dew forms when
    a) Air rises
    b) Air cools below the dew point
    c) Winds are strong
    d) Rain evaporates
22. Dew usually forms during
    a) Windy nights
    b) Humid days
    c) Clear, calm nights
    d) Rainy evenings
23. Frost forms when
    a) Temperature remains above freezing
    b) Water condenses
    c) Temperature falls below 0°C
    d) Rain occurs
24. Fog is a
    a) Form of precipitation
    b) Suspended water droplets near the ground
    c) Type of rainfall
    d) Result of evaporation
25. Fog reduces
    a) Wind speed
    b) Solar radiation
    c) Visibility
    d) Rainfall
26. Mist is different from fog because it
    a) Forms at night only
    b) Has larger water droplets
    c) Is denser than fog
    d) Has better visibility than fog
27. White frost is formed by
    a) Evaporation
    b) Freezing of dew
    c) Deposition of water vapor
    d) Cloud seeding
28. Which factor favors frost formation?
    a) Cloudy nights
    b) Windy nights
    c) Calm, clear nights
    d) Humid days

29. Which phenomenon involves deposition directly from vapor to ice?
    a) Condensation b) Evaporation
    c) Freezing d) Frost formation
30. The damage to crops by frost is mainly due to
    a) High humidity
    b) Ice crystals forming in plant tissues
    c) Soil freezing
    d) Heavy rain
31. Clouds are formed by
    a) Heating of air b) Rising & cooling of moist air
    c) Condensation of dew d) Solar radiation
32. Which of the following is a low-level cloud?
    a) Cirrus b) Stratus
    c) Altocumulus d) Cirrostratus
33. Cumulonimbus clouds are associated with
    a) Snowfall b) Thunderstorms
    c) Haze d) Calm weather
34. Cirrus clouds are found in:
    a) Troposphere base b) Lower atmosphere
    c) Upper troposphere d) Near the surface
35. Which cloud type gives continuous rain?
    a) Cirrus b) Altostratus
    c) Nimbostratus d) Cumulus
36. Which cloud is known for fair weather?
    a) Nimbostratus b) Cumulus
    c) Cirrostratus d) Cumulonimbus
37. Altostratus clouds are located in the
    a) High altitude b) Middle layer of troposphere
    c) Surface layer d) Polar region only
38. Which cloud appears feathery and white?
    a) Cumulus b) Stratus
    c) Cirrus d) Nimbostratus

39. Clouds are classified based on
    a) Height and appearance
    b) Color
    c) Region of formation
    d) Humidity
40. Lenticular clouds are shaped like
    a) Feather
    b) Mushroom
    c) Lens
    d) Column
41. High humidity is favorable for
    a) Transpiration
    b) Disease development
    c) Pest suppression
    d) Soil heating
42. Dew is beneficial for
    a) Disease spread
    b) Watering of crops
    c) Increasing wind speed
    d) None of the above
43. Which factor does not affect condensation?
    a) Temperature
    b) Air pressure
    c) Cloud color
    d) Humidity
44. Clouds affect
    a) Surface temperature
    b) Radiation balance
    c) Rainfall formation
    d) All of the above
45. Relative humidity can be increased by
    a) Decreasing temperature
    b) Increasing temperature
    c) Decreasing wind speed
    d) Increasing solar radiation
46. Frost is dangerous during
    a) Wheat sowing
    b) Fruit flowering
    c) Rice harvesting
    d) Soil preparation
47. What type of condensation forms without coming into contact with the ground?
    a) Dew
    b) Fog
    c) Mist
    d) Cloud
48. What is the main cause of early morning fog in winters?
    a) Evaporation
    b) Cold air and high humidity
    c) High wind
    d) Heavy rainfall
49. High cloud cover results in:
    a) More sunshine
    b) Lower night temperature
    c) Less radiation loss at night
    d) Faster dew formation

50. Cloud formation is crucial for:
   a) Photosynthesis b) Rainfall prediction
   c) Soil fertility d) Crop rotation

Answer Key

1	c	2	b	3	a	4	b	5	b	6	c	7	d
8	b	9	b	10	b	11	b	12	c	13	b	14	a
15	c	16	b	17	d	18	b	19	c	20	b	21	b
22	c	23	c	24	b	25	c	26	d	27	c	28	c
29	d	30	b	31	b	32	b	33	c	34	c	35	c
36	b	37	c	38	c	39	a	40	c	41	b	42	b
43	c	44	d	45	a	46	b	47	d	48	b	49	c
50	b												

Ch. 9. Cloud Formation and Classification

1. Clouds are formed due to the process of
   a) Condensation b) Evaporation
   c) Sublimation d) Radiation
2. The primary requirement for cloud formation is
   a) High pressure b) Condensation nuclei
   c) Wind speed d) Solar radiation
3. Which of the following is a condensation nucleus?
   a) Water vapor b) Dust particles
   c) Nitrogen gas d) Oxygen gas
4. Water changes from vapor to liquid during
   a) Evaporation b) Condensation
   c) Sublimation d) Transpiration
5. At what relative humidity do clouds generally begin to form?
   a) 50% b) 70%
   c) 100% d) 30%
6. Clouds are classified mainly based on
   a) Temperature and color b) Shape and altitude
   c) Latitude and longitude d) Pressure and wind
7. Which organization provides the international classification of clouds?
   a) FAO b) ICAR
   c) WMO d) IMD

8. Which of the following is a low-level cloud?
   a) Cirrus b) Stratus
   c) Cumulonimbus d) Altostratus
9. Cirrus clouds are found at:
   a) High altitudes b) Low altitudes
   c) Surface level d) Underground
10. Which clouds resemble feathery wisps?
   a) Cumulus b) Cirrus
   c) Stratus d) Nimbus
11. Clouds that appear like a flat, gray layer covering the sky are
   a) Stratus b) Cumulus
   c) Cirrus d) Altocumulus
12. Puffy, cotton-like clouds are called
   a) Cirrus b) Stratus
   c) Cumulus d) Nimbostratus
13. Altostratus and altocumulus are found in the
   a) Lower atmosphere b) Middle atmosphere
   c) Upper atmosphere d) Troposphere only
14. Which cloud type is typically associated with thunderstorms?
   a) Cirrostratus b) Cumulonimbus
   c) Altocumulus d) Stratus
15. Nimbus in cloud names indicates:
   a) High clouds b) Rain-bearing clouds
   c) Snow clouds d) Thunderstorm clouds
16. Cumulonimbus clouds can extend from:
   a) Surface to upper troposphere b) Troposphere to stratosphere
   c) Ground level to mesosphere d) Mid-troposphere only
17. Which cloud produces steady, prolonged rainfall?
   a) Cumulus b) Nimbostratus
   c) Cirrus d) Altocumulus
18. A cloud at ground level is called:
   a) Stratus b) Fog
   c) Mist d) Dew

19. What type of cloud usually indicates fair weather?
    a) Cumulus b) Nimbostratus
    c) Cumulonimbus d) Cirrostratus
20. The base of low clouds lies approximately at:
    a) 2 to 7 km b) 0 to 2 km
    c) 7 to 13 km d) 8 to 20 km
21. Instrument used to observe clouds is:
    a) Ceilometer b) Anemometer
    c) Hygrometer d) Pyranometer
22. Ceilometer is used to measure:
    a) Rainfall b) Wind speed
    c) Cloud base height d) Solar radiation
23. Which cloud type has vertical development?
    a) Stratus b) Cirrostratus
    c) Cumulonimbus d) Altostratus
24. A sky covered with altostratus clouds usually signals
    a) Dry weather b) Approaching storm or rain
    c) Snowfall d) Cold wave
25. Clouds that form in the middle altitude are
    a) Cirrus and cirrostratus b) Altostratus & altocumulus
    c) Cumulonimbus d) Fog
26. Supercooled water exists in clouds at temperatures
    a) Above 0°C b) Below 0°C
    c) At 100°C d) At -100°C
27. Bergeron process is associated with
    a) Fog formation b) Cloud seeding
    c) Ice crystal growth in clouds d) Radiation cooling
28. The key process that initiates rainfall from warm clouds is
    a) Radiation b) Collision-coalescence
    c) Sublimation d) Dispersion
29. Cloud droplets require a nucleus for condensation, called
    a) Aerosols b) CO_2
    c) Nitrogen d) Hydrogen

30. Which is not a method of cloud classification?
    a) Based on shape
    b) Based on height
    c) Based on temperature
    d) Based on precipitation potential
31. Clouds play a role in Earth's:
    a) Photosynthesis b) Carbon cycle
    c) Radiation balance d) Wind movement
32. Clouds reflect solar radiation, which leads to:
    a) Global warming b) Cooling effect
    c) Drought d) Greenhouse effect
33. Cloud cover reduces nighttime cooling by
    a) Conduction b) Radiative trapping
    c) Absorption d) Advection
34. Thick clouds have what effect on Earth's surface?
    a) Heating b) Cooling
    c) No effect d) Acidification
35. Clouds are part of which Earth system?
    a) Lithosphere b) Biosphere
    c) Atmosphere d) Cryosphere
36. Cloudy skies reduce
    a) Temperature b) Solar radiation
    c) Soil moisture d) Humidity
37. Which type of cloud is least favorable for drying of crops?
    a) Cirrus b) Stratus
    c) Altostratus d) Cumulus
38. Cloudy weather delays
    a) Harvesting b) Seed germination
    c) Maturity of crops d) Leaf formation
39. Cloud cover reduces potential for
    a) Crop growth b) Evapotranspiration
    c) Carbon sequestration d) Root respiration

40. Which season is associated with maximum cloud cover in India?
    a) Winter b) Summer
    c) Monsoon d) Autumn
41. What are mammatus clouds associated with?
    a) Calm weather b) Thunderstorms
    c) Hailstorms d) Drought
42. Lenticular clouds are lens-shaped and form due to:
    a) Mountain air flow b) Cyclone formation
    c) Sea breeze d) Cloudburst
43. Which clouds look like fish scales and indicate fair weather?
    a) Cirrocumulus b) Altostratus
    c) Nimbostratus d) Stratus
44. Which clouds are highest in the sky?
    a) Altostratus b) Cumulus
    c) Cirrus d) Nimbostratus
45. Pileus clouds form over
    a) Hills b) Volcanoes
    c) Cumulus towers d) Water bodies
46. Which is not a high-level cloud?
    a) Cirrus b) Cirrostratus
    c) Altocumulus d) Cirrocumulus
47. Cloud seeding is aimed at
    a) Removing pollution b) Artificial precipitation
    c) Blocking UV rays d) Reducing greenhouse gases
48. Clouds often form due to
    a) Descending air b) Rising air
    c) Horizontal wind d) Sinking cold fronts
49. Clouds help in estimation of
    a) Air composition b) Precipitation forecast
    c) Ozone depletion d) Soil moisture
50. Satellite images help monitor cloud
    a) Composition b) Temperature
    c) Coverage and movement d) Altitude only

Answer Key

1	a	2	b	3	b	4	b	5	c	6	b	7	c
8	b	9	a	10	b	11	a	12	c	13	b	14	b
15	b	16	a	17	b	18	b	19	a	20	b	21	a
22	c	23	c	24	b	25	b	26	b	27	c	28	b
29	a	30	c	31	c	32	b	33	b	34	b	35	c
36	b	37	b	38	c	39	b	40	c	41	b	42	a
43	a	44	c	45	c	46	c	47	b	48	b	49	b
50	c												

Ch. 10. Processes of condensation

1. Condensation is the process of change of water vapor into:
   a) Ice b) Snow
   c) Liquid water d) Steam
2. Condensation occurs when air is:
   a) Heated b) Saturated and cooled
   c) Pressurized d) Evaporated
3. Condensation is the reverse of:
   a) Melting b) Sublimation
   c) Evaporation d) Freezing
4. The temperature at which air becomes saturated and condensation begins is called:
   a) Boiling point b) Condensation index
   c) Dew point d) Latent point
5. Which of the following is essential for condensation to occur in the atmosphere?
   a) Wind b) Sunlight
   c) Condensation nuclei d) Vegetation
6. Condensation is favored when:
   a) Temperature rises b) Air is dry
   c) Humidity is high d) Wind speed is high
7. The common condensation nuclei include:
   a) Dust particles b) Nitrogen gas
   c) Hydrogen d) Oxygen

8. Which atmospheric condition is necessary for cloud formation?
   a) Warm and dry air b) Cool and saturated air
   c) Hot and unsaturated air d) Rapid heating of surface
9. Condensation occurs more readily when the air is:
   a) Clear and dry b) Calm and warm
   c) Cold and moist d) Warm and dry
10. Condensation takes place when relative humidity is:
   a) Less than 50% b) Around 80%
   c) 100% d) Zero
11. Dew is formed when condensation occurs
   a) In the sky b) On surface objects
   c) Inside clouds d) Under soil
12. Which of the following is not a form of condensation?
   a) Dew b) Fog
   c) Rain d) Frost
13. Mist differs from fog in
   a) Shape b) Visibility distance
   c) Formation process d) Chemical composition
14. Frost forms when condensation happens
   a) Below 0°C b) Above 30°C
   c) In sunlight d) At high wind speed
15. Which form of condensation appears like a cloud near the ground?
   a) Dew b) Fog
   c) Rain d) Drizzle
16. Dew forms most commonly during
   a) Rainy nights b) Cloudy nights
   c) Clear, calm nights d) Stormy nights
17. Which condition is not conducive to dew formation?
   a) Clear skies b) Calm winds
   c) High wind speed d) High humidity
18. Frost is common when the sky is
   a) Cloudy b) Clear
   c) Hazy d) Rainy

19. Fog can disrupt
    a) Photosynthesis b) Radiation
    c) Transportation d) Germination
20. Which of the following condensation forms is solid?
    a) Mist b) Fog
    c) Dew d) Frost
21. Condensation releases
    a) Heat energy b) Cold energy
    c) Potential energy d) Electrical energy
22. The latent heat of condensation helps in
    a) Heating the air mass b) Cooling the soil
    c) Freezing water d) Condensing CO_2
23. Clouds form when water vapor condenses around
    a) Ice particles b) CO_2
    c) Dust or smoke d) Rain drops
24. In agrometeorology, condensation indicates potential
    a) Earthquake b) Rainfall
    c) Heatwave d) Drought
25. High relative humidity and cooling below dew point result in
    a) Heating b) Fog formation
    c) Evaporation d) Solar radiation
26. Dew is beneficial for crops in
    a) Arid areas b) Flooded fields
    c) Rainfed rice d) Saline soils
27. Frost can cause damage to
    a) Mature crops b) Germinating seeds
    c) Flowering and fruiting stages d) Roots only
28. Fog reduces
    a) Humidity b) Solar radiation
    c) Dew point d) Soil fertility

29. Dew contributes to
    a) Disease in crops
    b) Additional moisture availability
    c) Reduced photosynthesis
    d) Hailstorms
30. Frost injury is more common in
    a) Paddy
    b) Wheat
    c) Maize
    d) Sugarcane
31. Condensation depends on
    a) Soil type
    b) Temperature and relative humidity
    c) Fertility
    d) Rainfall pattern
32. Dew point is measured in
    a) %
    b) mm
    c) °C
    d) Lux
33. Frost may damage crops due to
    a) High radiation
    b) Cell bursting from ice crystal formation
    c) Increased respiration
    d) Evaporation
34. Mist is formed when
    a) Air is hot and dry
    b) Dew evaporates
    c) Condensation occurs in warm air with slight cooling
    d) Ice melts
35. Fog and mist reduce
    a) Transpiration
    b) Visibility
    c) Photosynthesis
    d) Fertility
36. In the upper atmosphere, condensation leads to:
    a) Soil heating
    b) Cloud formation
    c) Wind blowing
    d) Volcanic eruption

37. Saturated air has a relative humidity of:
    a) 0%
    b) 50%
    c) 100%
    d) 150%
38. Which weather condition involves condensation at the surface level?
    a) Haze
    b) Dew
    c) Drizzle
    d) Snow
39. Condensation is an example of a
    a) Physical process
    b) Chemical process
    c) Nuclear process
    d) Geological process
40. Which instrument measures dew point?
    a) Anemometer
    b) Dew cell or hygrometer
    c) Thermometer
    d) Rain gauge
41. Frost-free days are important for
    a) Flowering crops
    b) Soil testing
    c) Irrigation
    d) Weeding
42. Radiation frost occurs due to:
    a) Cloudy nights
    b) High humidity
    c) Rapid cooling of Earth's surface on clear nights
    d) Rainfall
43. Which crop is most sensitive to frost?
    a) Sorghum
    b) Mustard
    c) Potato
    d) Bajra
44. Condensation nuclei are essential for:
    a) Wind movement
    b) Cloud development
    c) Hail formation
    d) Dew evaporation
45. Which layer of the atmosphere is condensation most likely to occur?
    a) Stratosphere
    b) Troposphere
    c) Mesosphere
    d) Thermosphere
46. Condensation releases energy in the form of:
    a) Kinetic energy
    b) Latent heat
    c) Magnetic energy
    d) Potential energy

47. Condensation cannot occur unless air is:
    a) Supercooled b) Saturated
    c) Heated d) Diluted
48. Which part of the day is dew most likely to form?
    a) Afternoon b) Midnight
    c) Early morning d) Noon
49. The energy released during condensation contributes to:
    a) Soil erosion
    b) Cloud growth and storm intensification
    c) Evaporation
    d) Plant metabolism
50. If the air is below freezing and condensation occurs, the result is:
    a) Fog b) Frost
    c) Dew d) Rain

Answer Key

1	c	2	b	3	c	4	c	5	c	6	c	7	a
8	b	9	c	10	c	11	b	12	c	13	b	14	a
15	b	16	b	17	c	18	b	19	c	20	d	21	a
22	a	23	c	24	b	25	b	26	a	27	c	28	b
29	b	30	b	31	b	32	c	33	b	34	c	35	b
36	b	37	c	38	b	39	a	40	b	41	a	42	c
43	c	44	b	45	b	46	b	47	b	48	c	49	b
50	b												

Ch.11. Precipitation & its Forms

1. Precipitation refers to
    a) Solar radiation b) Water released from clouds
    c) Air pressure increase d) Wind formation
2. The primary requirement for precipitation is
    a) Wind shear
    b) Evaporation
    c) Condensation and cloud formation
    d) High temperature

3. The key process responsible for rainfall is
   a) Evaporation b) Precipitation
   c) Condensation d) Transpiration
4. Which of the following is not a form of precipitation?
   a) Rain b) Fog
   c) Snow d) Hail
5. The coalescence process in clouds leads to:
   a) Cloud dissipation b) Raindrop formation
   c) Hail formation d) Wind movement
6. Bergeron process operates mainly in:
   a) Tropical clouds b) Warm clouds
   c) Cold clouds d) Surface water
7. What causes water droplets in clouds to fall as rain?
   a) Solar radiation
   b) Decreased air pressure
   c) Droplet growth beyond fall velocity threshold
   d) Evaporation of droplets
8. Cloud droplets typically collide and merge in:
   a) Bergeron process b) Coalescence process
   c) Sublimation d) Evaporation
9. Cloud droplets must be at least ______ to fall as rain:
   a) 0.01 mm b) 0.1 mm
   c) 1.0 mm d) 5.0 mm
10. Virga is a condition when:
    a) Hail falls on the ground
    b) Snow melts before reaching ground
    c) Rain evaporates before reaching ground
    d) Cloud evaporates completely
11. Rainfall is classified as precipitation in the form of:
    a) Ice crystals b) Water droplets > 0.5 mm
    c) Snowflakes d) Water vapor
12. Drizzle consists of:
    a) Snow b) Water droplets < 0.5 mm
    c) Frozen pellets d) Fog particles

13. Snow is precipitation in the form of:
    a) Frozen raindrops
    b) Ice needles
    c) Ice crystals forming hexagonal flakes
    d) Hailstones
14. Sleet refers to:
    a) Ice crystals
    b) Rain mixed with snow
    c) Frozen rain droplets
    d) Heavy rainfall
15. Hail is formed when:
    a) Water condenses
    b) Snow accumulates
    c) Water droplets freeze in strong updrafts
    d) Evaporation cools air
16. Which precipitation is most damaging to crops?
    a) Rain
    b) Snow
    c) Hail
    d) Drizzle
17. Which of the following is not a frozen form of precipitation?
    a) Snow
    b) Sleet
    c) Hail
    d) Drizzle
18. Freezing rain occurs when
    a) Rain falls on hot ground
    b) Rain freezes before reaching ground
    c) Rain freezes on contact with cold surfaces
    d) Snow melts on falling
19. The typical diameter of hailstones is
    a) <1 mm
    b) 1–5 mm
    c) 5–50 mm
    d) 1 meter
20. Which type of precipitation is associated with cumulonimbus clouds?
    a) Snow
    b) Drizzle
    c) Hail
    d) Fog
21. Clouds are formed by
    a) Warming of moist air
    b) Expansion and cooling of air
    c) Sublimation
    d) Reflection of sunlight

22. Condensation nuclei are essential for
    a) Rainfall b) Cloud formation
    c) Wind generation d) Lightning
23. Cumulus clouds are characterized by
    a) Flat layers b) Wispy structures
    c) Puffy appearance d) Dark bottom layers
24. Stratus clouds typically form at
    a) High altitudes b) Low altitudes
    c) Polar regions d) Equator
25. Cirrus clouds appear as
    a) Sheet-like b) Wispy and feathery
    c) Dark and dense d) Puffy
26. Nimbostratus clouds bring
    a) Fair weather b) Strong winds
    c) Continuous rainfall d) No precipitation
27. Which cloud is associated with thunderstorms?
    a) Cumulus b) Stratus
    c) Nimbostratus d) Cumulonimbus
28. Which cloud classification is based on altitude and shape?
    a) Köppen classification b) International Cloud Atlas
    c) Bergeron classification d) Albedo method
29. Which of the following is a high-level cloud?
    a) Cirrus b) Altostratus
    c) Nimbostratus d) Stratus
30. Cloud base height is typically measured using:
    a) Thermograph b) Ceilometer
    c) Hygrometer d) Barometer
31. Artificial rainmaking is also known as:
    a) Evaporation b) Cloud suppression
    c) Cloud seeding d) Fog dispersal
32. The main aim of cloud seeding is to:
    a) Prevent clouds b) Increase wind speed
    c) Induce or enhance precipitation d) Stop evaporation

33. Which of the following is used in cloud seeding?
    a) Sodium chloride b) Silver iodide
    c) Ammonium nitrate d) Urea
34. Cloud seeding promotes:
    a) Evaporation b) Condensation
    c) Wind flow d) Thunderstorms
35. Which type of clouds are most suitable for seeding?
    a) Cirrus b) Stratus
    c) Cumulonimbus d) Lenticular
36. Artificial rainmaking was first tested in:
    a) India b) USA
    c) Russia d) China
37. Most common method of cloud seeding is:
    a) Sonic seeding b) Rocket launching
    c) Aircraft dispersion d) Infrared radiation
38. Cloud seeding can be used to suppress
    a) Hail formation b) Droughts
    c) Flooding d) Cyclones
39. Which of the following is not used in cloud seeding?
    a) Potassium chloride b) Dry ice
    c) Silver iodide d) Nitrogen
40. Cloud seeding is less effective when
    a) Humidity is high b) Natural clouds are present
    c) Clouds are too warm or too dry d) Winds are calm
41. Rainfall intensity is measured in
    a) mm per hour b) mm per day
    c) cm per second d) Litre per square meter
42. Total rainfall is measured by
    a) Hygrometer b) Rain gauge
    c) Barometer d) Ceilometer
43. In India, rainfall is mainly due to
    a) Westerlies b) Cyclones
    c) Monsoon winds d) Local storms

44. The most efficient rainmaking agent is
    a) Ice b) Salt
    c) Silver iodide d) Water vapor
45. Which region in India uses cloud seeding to mitigate droughts?
    a) Rajasthan b) Kerala
    c) Assam d) Bihar
46. Heavy precipitation in a short time is called
    a) Showers b) Drizzle
    c) Cloudburst d) Rainstorm
47. Precipitation in freezing conditions often forms
    a) Rain b) Sleet or snow
    c) Fog d) Dew
48. Fog is different from precipitation because
    a) It forms above the ground b) It doesn't fall
    c) It is frozen d) It is warmer
49. Excessive hail can damage:
    a) Soil texture b) Photosynthesis
    c) Crop yield d) All of the above
50. Which form of precipitation is essential for rabi crops in North India?
    a) Snow b) Monsoon rain
    c) Western disturbance rain d) Artificial rain

Answer Key

1	b	2	c	3	c	4	b	5	b	6	c	7	c
8	b	9	c	10	c	11	b	12	b	13	c	14	c
15	c	16	c	17	d	18	c	19	c	20	c	21	b
22	b	23	c	24	b	25	b	26	c	27	d	28	b
29	a	30	b	31	c	32	c	33	b	34	b	35	c
36	b	37	c	38	a	39	d	40	c	41	a	42	b
43	c	44	c	45	a	46	c	47	b	48	b	49	d
50	c												

Ch. 12. Monsoons & its Mechanism

1. What is the primary purpose of artificial rainmaking?
   a) To prevent rain
   b) To increase rainfall
   c) To measure cloud height
   d) To study cloud colors
2. Which of the following chemicals is most commonly used for cloud seeding?
   a) Sodium chloride
   b) Silver iodide
   c) Calcium carbonate
   d) Potassium nitrate
3. Artificial rainmaking is also known as
   a) Cloud bursting
   b) Cloud harvesting
   c) Cloud seeding
   d) Rain suppression
4. Which agency first successfully carried out cloud seeding in India?
   a) ISRO
   b) IMD
   c) IITM, Pune
   d) DRDO
5. Which aircraft is commonly used for cloud seeding operations in India?
   a) Sukhoi Su-30
   b) Dornier Do-228
   c) Boeing 747
   d) Mirage 2000
6. Which of the following is NOT a method of cloud seeding?
   a) Ground-based generators
   b) Aircraft dispersion
   c) Balloon release
   d) Satellite laser beams
7. In artificial rainmaking, hygroscopic substances such as ___ are used for warm cloud seeding.
   a) Silver iodide
   b) Dry ice
   c) Sodium chloride
   d) Calcium sulfate
8. Which one is a limitation of artificial rainmaking?
   a) Increases cloud cover
   b) Only works in winter
   c) Depends on cloud availability
   d) Can cause snowfall
9. What is the role of seeding agents in cloud seeding?
   a) To disperse the clouds
   b) To suppress condensation
   c) To initiate nucleation
   d) To evaporate water
10. Which state in India has undertaken extensive cloud seeding programs?
   a) Punjab
   b) Maharashtra
   c) Kerala
   d) Odisha

11. Cold cloud seeding is most effective in clouds with temperatures below:
    a) 0°C b) 10°C
    c) 4°C d) -5°C
12. Which seeding agent is typically used in cold cloud seeding?
    a) Silver iodide b) Sodium chloride
    c) Ammonium nitrate d) Sulfur dioxide
13. The Bergeron process is associated with:
    a) Warm cloud seeding b) Cold cloud formation
    c) Cloud dissipation d) Artificial lightning
14. Warm cloud seeding primarily aims at enhancing:
    a) Ice crystal formation b) Snowfall
    c) Coalescence of water droplets d) Freezing of raindrops
15. Which of the following best describes warm cloud seeding?
    a) Uses chemicals to cool the clouds
    b) Involves dry ice and silver iodide
    c) Promotes growth of raindrops in liquid water clouds
    d) Only effective during the winter
16. Cold cloud seeding leads to rainfall by
    a) Ice crystal growth
    b) Coalescence of water drops
    c) Evaporation of small drops
    d) Condensation on dust particles
17. Which is NOT a characteristic of cold clouds?
    a) Contain supercooled water b) Found above freezing level
    c) Contain only water vapor d) May include ice crystals
18. Dry ice used in cloud seeding works by
    a) Heating the cloud
    b) Cooling the cloud to promote freezing
    c) Dissolving cloud particles
    d) Absorbing water vapor
19. Cold cloud seeding requires clouds to be at what altitude typically?
    a) Below 1000 m b) 2000–3000 m
    c) Above 5000 m d) Near sea level

20. Silver iodide works effectively because it has a crystal structure similar to
    a) Water vapor b) Ice
    c) Salt d) Oxygen
21. Indian monsoon is primarily caused due to
    a) Ocean currents
    b) Rotation of Earth
    c) Differential heating of land and sea
    d) Solar flares
22. The monsoon in India generally arrives by
    a) February b) May
    c) June d) August
23. Which ocean plays a major role in the Indian monsoon?
    a) Atlantic Ocean b) Pacific Ocean
    c) Arctic Ocean d) Indian Ocean
24. The ITCZ (Inter-Tropical Convergence Zone) shifts northward during:
    a) Summer b) Winter
    c) Monsoon withdrawal d) Spring
25. Which branch of monsoon strikes Kerala coast first?
    a) Bay of Bengal b) Arabian Sea
    c) Himalayan d) Tibetan
26. The El Niño phenomenon generally causes
    a) Enhanced Indian monsoon b) Suppressed Indian monsoon
    c) No impact d) Snowfall in monsoon
27. Which of the following is responsible for breaking monsoon in India?
    a) Western disturbances b) Madden-Julian Oscillation
    c) Cold fronts d) Sea breeze
28. Retreating monsoon in India occurs in
    a) June–July b) August–September
    c) October–November d) December–January
29. Monsoon trough is a
    a) High-pressure zone b) Low-pressure area
    c) Jet stream d) Cyclonic eye

30. Monsoon onset in India is declared by
    a) Indian Army
    b) Indian Meteorological Department
    c) Ministry of Agriculture
    d) World Bank
31. What percentage of Indian agriculture depends on monsoon rains?
    a) 20% b) 40%
    c) 60% d) 75%
32. Deficient monsoon in India leads to
    a) Improved yields b) Floods
    c) Droughts d) Salinity
33. Which crop is most vulnerable to monsoon failure?
    a) Wheat b) Maize
    c) Rice d) Barley
34. Kharif crops in India are sown in
    a) October b) January
    c) June-July d) March
35. Which of the following is a Kharif crop dependent on monsoon?
    a) Mustard b) Wheat
    c) Groundnut d) Gram
36. Monsoon variability impacts
    a) Crop selection b) Sowing time
    c) Yield d) All of the above
37. Which irrigation method becomes crucial in poor monsoon years?
    a) Flood irrigation b) Drip irrigation
    c) Overhead irrigation d) Tank irrigation
38. Excessive monsoon rains can cause
    a) Soil erosion b) Increased yields
    c) Frost d) Cloud seeding
39. Which agro-meteorological service is used for monsoon monitoring?
    a) Agro-forestry mapping
    b) FASAL
    c) Gramin Krishi Mausam Sewa (GKMS)
    d) APEDA

40. Rainfed farming in India is practiced mostly during:
    a) Rabi season
    b) Kharif season
    c) Zaid season
    d) Winter
41. In India, cloud seeding is permitted under which environmental act?
    a) Water Act 1974
    b) Environmental Protection Act 1986
    c) There is no specific law
    d) Air Act 1981
42. Which is the best time for cloud seeding in India?
    a) Before monsoon
    b) Peak monsoon
    c) Post monsoon
    d) Winter
43. Which parameter is crucial to assess cloud seeding potential?
    a) Wind speed
    b) Cloud temperature and moisture
    c) Soil pH
    d) Atmospheric pressure
44. Monsoon prediction is currently done using:
    a) Animal behavior
    b) Satellite & climate models
    c) Folk sayings
    d) Sea wave patterns
45. What is the role of aerosols in cloud formation?
    a) Prevent rain
    b) Serve as cloud condensation nuclei
    c) Evaporate raindrops
    d) Reflect sunlight
46. Which satellite is used for Indian monsoon monitoring?
    a) INSAT-3D
    b) Kalpana-1
    c) GSAT-6
    d) Aryabhata
47. Which Indian institution is at the forefront of cloud microphysics research?
    a) IIT Delhi
    b) IARI
    c) IISc Bangalore
    d) IITM Pune
48. Which is more uncertain—cold or warm cloud seeding?
    a) Cold
    b) Warm
    c) Both are equally certain
    d) Neither

49. A successful monsoon improves which of the following?
 a) Food security b) Farm employment
 c) GDP growth d) All of the above
50. Which is the most studied weather modification method in India?
 a) Lightning redirection b) Hailstorm suppression
 c) Cloud seeding d) Dew harvesting

Answer Key

1	b	2	b	3	c	4	c	5	b	6	d	7	c
8	c	9	c	10	b	11	a	12	a	13	b	14	c
15	c	16	a	17	c	18	b	19	c	20	b	21	c
22	c	23	d	24	a	25	b	26	b	27	b	28	c
29	b	30	b	31	d	32	c	33	c	34	c	35	c
36	d	37	b	38	a	39	c	40	b	41	c	42	a
43	b	44	b	45	b	46	a	47	d	48	b	49	d
50	c												

Ch. 13. Weather hazards

1. Drought is defined as a condition of:
 a) Excess rainfall b) Above-normal humidity
 c) Prolonged dry weather d) Excess soil moisture
2. Which type of drought is associated with insufficient rainfall?
 a) Meteorological drought b) Agricultural drought
 c) Hydrological drought d) Socio-economic drought
3. Agricultural drought primarily affects:
 a) Rivers b) Lakes
 c) Crop production d) Industry
4. Hydrological drought relates to:
 a) Surface and subsurface water deficit
 b) Atmospheric dryness
 c) Seed dormancy
 d) Soil type
5. Which Indian state is most prone to recurrent droughts?
 a) Punjab b) Kerala
 c) Rajasthan d) Assam

6. Which index is commonly used for drought monitoring?
   a) Wind Chill Index
   b) Standardized Precipitation Index (SPI)
   c) Humidity Index
   d) Cloud Cover Index
7. Which crop is more drought-tolerant?
   a) Rice b) Maize
   c) Sorghum d) Sugarcane
8. Drought-resistant varieties are part of:
   a) Climate mitigation b) Genetic erosion
   c) Adaptive farming d) Soil salinization
9. Which farming method helps reduce drought impact?
   a) Flood irrigation b) Rainwater harvesting
   c) Mono-cropping d) Shifting cultivation
10. Which agency monitors drought conditions in India?
   a) NABARD b) IMD
   c) NDMA d) NITI Aayog
11. Floods occur due to:
   a) Lack of clouds
   b) Excess rainfall and poor drainage
   c) Excessive evaporation
   d) Heat waves
12. Flash floods are usually caused by:
   a) Gradual snowmelt b) Sudden intense rainfall
   c) High humidity d) Temperature rise
13. Which Indian region is frequently affected by floods?
   a) Western Ghats b) Himalayan region
   c) Indo-Gangetic plains d) Thar Desert
14. Which crop is most vulnerable to floods?
   a) Rice b) Cotton
   c) Sugarcane d) Pulses
15. Waterlogging leads to
   a) Improved aeration b) Root suffocation
   c) Enhanced nutrient uptake d) Higher photosynthesis

16. Which is a major consequence of floods on agriculture?
    a) Better root development
    b) Enhanced pollination
    c) Soil erosion and crop damage
    d) Improved seed germination
17. Which structure helps manage agricultural flood risks?
    a) Check dams
    b) Wind turbines
    c) Pesticide sprayers
    d) Greenhouses
18. Which river causes frequent flooding in Assam?
    a) Yamuna
    b) Krishna
    c) Brahmaputra
    d) Godavari
19. Flood-tolerant rice variety developed in India:
    a) IR 64
    b) Swarna-Sub1
    c) Pusa Basmati
    d) Sona Masuri
20. Which satellite helps monitor flood conditions?
    a) INSAT-3D
    b) RISAT
    c) Aryabhata
    d) GSAT-6
21. Frost is a weather hazard caused by
    a) High temperatures
    b) Rapid warming
    c) Air temperature below freezing point
    d) Dew formation
22. Frost typically occurs during
    a) Summer nights
    b) Cloudy nights
    c) Clear and calm winter nights
    d) Monsoon days
23. Frost damage is common in:
    a) Sugarcane
    b) Tea and coffee
    c) Bajra
    d) Paddy
24. Which of the following is not a type of frost?
    a) Radiation frost
    b) Advection frost
    c) Convective frost
    d) Hoar frost
25. Frost damages plants by
    a) Melting roots
    b) Destroying cell structure due to ice crystals
    c) Enhancing photosynthesis
    d) Increasing respiration

26. Which method helps in frost protection?
    a) Flooding fields b) Using wind machines
    c) Removing mulch d) Burning crop residues
27. Sensitive growth stage for frost in crops is
    a) Vegetative b) Germination
    c) Flowering d) Harvest
28. Which chemical is used in frost protection?
    a) Silver iodide b) Anti-transpirants
    c) Cryoprotectants d) Insecticides
29. Which region in India experiences frost hazards regularly?
    a) Sunderbans b) Western Rajasthan
    c) North Indian plains d) Coastal Andhra
30. Which fruit crop is highly sensitive to frost?
    a) Banana b) Mango
    c) Apple d) Papaya
31. Tropical cyclones form over:
    a) Mountains b) Hot deserts
    c) Warm ocean waters d) Arctic region
32. Cyclone intensity is measured by:
    a) Richter scale b) Saffir-Simpson scale
    c) Fujita scale d) Moh scale
33. Cyclones affecting India usually originate in:
    a) Bay of Bengal and Arabian Sea b) Red Sea
    c) Dead Sea d) Caspian Sea
34. Cyclones cause damage through:
    a) Heavy rainfall b) Strong winds
    c) Storm surges d) All of the above
35. Which state is most vulnerable to cyclones in India?
    a) Gujarat b) West Bengal
    c) Odisha d) Haryana
36. Tropical cyclones are known as hurricanes in:
    a) Japan b) USA
    c) India d) Africa

37. The eye of a tropical cyclone is
    a) The coldest region
    b) The calm and clear area
    c) The area of highest wind speed
    d) The warmest and windiest
38. Cyclone early warning in India is managed by
    a) FCI
    b) ICAR
    c) IMD
    d) CWC
39. Post-cyclone farming damage includes
    a) Lodging
    b) Waterlogging
    c) Salinity intrusion
    d) All of the above
40. Which is an example of a major Indian cyclone?
    a) Tauktae
    b) Phailin
    c) Amphan
    d) All of the above
41. A heat-wave is declared in India when temperature exceeds:
    a) 30°C
    b) 35°C
    c) 40°C (with deviation from normal)
    d) 25°C
42. Cold-waves are associated with
    a) Increased rainfall
    b) Sharp drop in night temperatures
    c) Thunderstorms
    d) Fog formation only
43. Which crop is highly sensitive to heat stress at flowering?
    a) Wheat
    b) Groundnut
    c) Sorghum
    d) Cotton
44. Heat-waves commonly affect which region of India?
    a) Coastal Karnataka
    b) North-west India
    c) Kerala
    d) Northeast India
45. Cold-waves affect agriculture by:
    a) Delaying flowering
    b) Reducing germination
    c) Causing frost
    d) All of the above
46. Which is not a mitigation method for heat-wave in crops?
    a) Light irrigation
    b) White mulching
    c) Delayed sowing
    d) Shade netting

47. Heat stress in livestock leads to:
    a) Increased feed intake b) Higher milk yield
    c) Reduced productivity d) Faster growth
48. Which forecast is crucial for heat-wave warnings?
    a) Rainfall forecast b) Temperature forecast
    c) Wind forecast d) Humidity forecast
49. Cold stress in plants can be reduced by:
    a) Water stress b) Spray of micronutrients
    c) Smoking near fields d) Frequent irrigation
50. Extreme weather events are increasing due to:
    a) Seasonal variation b) Natural disasters
    c) Climate change d) Earth's magnetic field

Answer Key

1	c	2	a	3	c	4	a	5	c	6	b	7	c
8	c	9	b	10	b	11	b	12	b	13	c	14	d
15	b	16	c	17	a	18	c	19	b	20	b	21	c
22	c	23	b	24	c	25	b	26	b	27	c	28	c
29	c	30	d	31	c	32	b	33	a	34	d	35	c
36	b	37	b	38	c	39	d	40	d	41	c	42	b
43	a	44	b	45	d	46	c	47	c	48	b	49	c
50	c												

Ch. 14. Microclimatic Modification

1. Microclimate refers to the climate of
    a) Entire region b) Large continent
    c) Small or specific area d) Upper atmosphere
2. Which of the following does NOT influence microclimate?
    a) Soil type b) Vegetation
    c) Tectonic plates d) Topography
3. Microclimate is important in agriculture because it affects
    a) Political systems b) Crop growth directly
    c) Ocean currents d) Animal migration
4. The boundary layer is the layer of air
    a) In the ionosphere b) Close to the Earth's surface
    c) In the thermosphere d) Outside the atmosphere

5. Which crop benefits most from modified microclimate practices like mulching?
   a) Wheat b) Tomato
   c) Paddy d) Maize
6. Major elements of microclimate include all except:
   a) Temperature b) Humidity
   c) Latitude d) Wind
7. Microclimate can be controlled by manipulating:
   a) Soil only b) Vegetation only
   c) Local environmental conditions d) Global temperatures
8. A dense crop canopy modifies microclimate by reducing:
   a) Soil erosion
   b) Soil temperature and wind speed
   c) Cloud formation
   d) Relative humidity
9. Which type of soil has greater impact on microclimate?
   a) Sandy b) Clay
   c) Loam d) Peaty
10. Radiation balance plays a key role in modifying:
    a) Photosynthesis b) Crop yield
    c) Microclimate d) Rainfall
11. Which practice increases soil temperature?
    a) Black plastic mulch b) Green mulch
    c) White polyethylene d) Shade nets
12. Which type of mulch helps cool the soil?
    a) Black b) Organic
    c) Transparent d) Aluminum foil
13. Planting direction that minimizes wind speed is:
    a) East-West b) North-South
    c) Diagonal d) Irregular
14. Shelterbelts are primarily used to reduce:
    a) Soil moisture b) Wind speed
    c) Soil pH d) Fertilizer loss

15. Altering row spacing affects
    a) Soil fertility
    b) Microclimate through air circulation
    c) Rainfall
    d) Soil salinity
16. Shade nets are commonly used to reduce
    a) Wind velocity
    b) Radiation and temperature
    c) Soil nutrients
    d) Air pressure
17. Low tunnels are effective in
    a) Enhancing rainfall
    b) Protecting crops from frost
    c) Generating electricity
    d) Reducing water stress
18. Greenhouses modify microclimate by
    a) Increasing CO_2 only
    b) Increasing temperature and humidity
    c) Reducing light
    d) Freezing air
19. Polyhouses are commonly used for
    a) Rice cultivation
    b) Vegetables and floriculture
    c) Forest crops
    d) Rainfed crops
20. The primary purpose of micro-irrigation in modifying microclimate is:
    a) To enhance evaporation
    b) To reduce transpiration loss
    c) To supply water efficiently and regulate moisture
    d) To leach salts
21. Windbreaks reduce wind speed over a distance of
    a) 1x tree height
    b) 5x tree height
    c) 5–10x tree height
    d) 20x tree height
22. Best tree species for windbreaks should be
    a) Shallow rooted
    b) Tall, fast-growing
    c) Fruit-bearing
    d) Seasonal
23. The maximum reduction in wind speed is observed on:
    a) Windward side
    b) Leeward side
    c) Both sides equally
    d) Not influenced

24. Effective windbreaks have
    a) Only one tree row
    b) Multiple species with gaps
    c) Dense multiple rows with staggered planting
    d) Random planting
25. Shelterbelts contribute to microclimate by
    a) Enhancing photosynthesis
    b) Increasing surface runoff
    c) Reducing evapotranspiration and soil erosion
    d) Reducing seed germination
26. Intercropping helps in
    a) Lowering photosynthesis
    b) Soil hardening
    c) Modifying microclimate through canopy interactions
    d) Reducing biodiversity
27. Closer crop spacing leads to
    a) Increased sunlight to soil   b) Higher weed growth
    c) Increased canopy coverage   d) Soil erosion
28. Leaf area index (LAI) influences
    a) Soil fertility
    b) Microclimate and transpiration rate
    c) Irrigation schedule
    d) Soil type
29. Canopy management in orchards affects
    a) Global climate   b) Wind patterns
    c) Local microclimate   d) Water table
30. Thinning in crops is done to
    a) Increase competition
    b) Reduce evaporation
    c) Improve light penetration and airflow
    d) Increase pests

31. Soil mulching helps in
    a) Increasing pests
    b) Decreasing humidity
    c) Conserving soil moisture and moderating temperature
    d) Reducing organic matter
32. Drip irrigation modifies microclimate by
    a) Flooding the field
    b) Cooling the atmosphere
    c) Localized wetting of soil and maintaining RH
    d) Increasing wind speed
33. Water bodies near a field influence microclimate by
    a) Reducing humidity
    b) Increasing temperature
    c) Moderating temperature and humidity
    d) Enhancing frost
34. Raised beds and ridges affect
    a) Soil fertility
    b) Wind erosion
    c) Soil moisture and root zone temperature
    d) Solar radiation
35. Puddling in rice fields helps
    a) Increase soil porosity
    b) Reduce microclimate effect
    c) Create anaerobic conditions and reduce temperature variability
    d) Improve light penetration
36. Radiation reflectance is increased by
    a) Black plastic   b) White or silver mulch
    c) Green nets   d) Bare soil
37. Which tool is used to monitor soil microclimate?
    a) Pyranometer   b) Infrared thermometer
    c) Rain gauge   d) Wind vane

38. Evaporative cooling is useful in
    a) Raising temperature
    b) Reducing humidity
    c) Controlling greenhouse microclimate
    d) Increasing CO_2
39. Soil color influences microclimate by altering
    a) Wind speed
    b) Temperature absorption and retention
    c) Rainfall
    d) Evaporation
40. Surface albedo refers to
    a) Temperature of soil
    b) Ratio of reflected to incoming solar radiation
    c) Soil fertility
    d) Soil porosity
41. Climate-smart agriculture emphasizes:
    a) Global warming
    b) Reducing microclimate effects
    c) Adapting and mitigating microclimatic changes
    d) Monocropping
42. Climate change affects microclimates by
    a) Increasing uniformity
    b) Reducing weather extremes
    c) Increasing variability and unpredictability
    d) Making rainfall reliable
43. Agroforestry helps in:
    a) Reducing yields
    b) Enhancing microclimate regulation and carbon sequestration
    c) Reducing biodiversity
    d) Evaporating soil moisture
44. Cover crops are grown to
    a) Increase runoff
    b) Modify temperature and reduce erosion
    c) Reduce organic matter
    d) Reflect solar radiation

45. Windbreaks are most useful in
    a) Forest areas
    b) Coastal zones
    c) Arid and semi-arid zones
    d) Urban parks
46. Microclimate monitoring in precision farming is done using:
    a) Soil augers
    b) IoT sensors & weather stations
    c) Sprayers
    d) Manual charts
47. Changing plant density affects microclimate through:
    a) Soil pH
    b) Airflow and humidity
    c) Carbonates
    d) Nutrient levels
48. Vertical farming is an advanced method to control
    a) Regional climate
    b) Soil salinity
    c) Microclimate in indoor farming systems
    d) Seed quality
49. Which method is best for smallholder farmers to manage microclimate?
    a) GIS mapping
    b) Large greenhouses
    c) Mulching and crop spacing
    d) Hydroponics
50. Microclimate modification is most useful in
    a) Mechanized farming
    b) Protected cultivation and sensitive crops
    c) Agro-industrial units
    d) Forest management

Answer Key

1	c	2	c	3	b	4	b	5	b	6	c	7	c
8	b	9	a	10	c	11	a	12	b	13	b	14	b
15	b	16	b	17	b	18	b	19	b	20	c	21	c
22	b	23	b	24	c	25	c	26	c	27	c	28	b
29	c	30	c	31	c	32	c	33	c	34	c	35	c
36	b	37	b	38	c	39	b	40	b	41	c	42	c
43	b	44	b	45	c	46	b	47	b	48	c	49	c
50	b												

Ch. 15. Climatic Normals for Crops and Livestock

1. Climatic normals refer to:
   a) Average crop yield
   b) Long-term average of meteorological parameters
   c) Daily weather readings
   d) Cloud patterns
2. Climatic normals are usually calculated over a period of:
   a) 5 years b) 10 years
   c) 30 years d) 50 years
3. Agroclimatic classification is primarily based on:
   a) Latitude and longitude b) Soil type
   c) Rainfall and temperature d) Crop rotation
4. Optimum temperature for C3 crop photosynthesis is:
   a) 25–30°C b) 10–15°C
   c) 40–45°C d) 5–10°C
5. C4 plants like maize grow best in temperatures of:
   a) 10–20°C b) 25–35°C
   c) 35–45°C d) 15–25°C
6. Wheat requires optimal temperature of:
   a) 35–40°C b) 20–25°C
   c) 10–15°C d) 5–10°C
7. Wheat is classified as a:
   a) Tropical crop b) Temperate crop
   c) Arid crop d) Aquatic crop
8. Paddy (rice) requires average rainfall of:
   a) 100–150 mm b) 500–700 mm
   c) 1000–1500 mm d) Over 3000 mm
9. The critical temperature range for rice flowering is:
   a) 35–40°C b) 25–30°C
   c) 15–20°C d) 10–15°C
10. Maize grows best at temperatures of:
    a) 18–22°C b) 30–35°C
    c) 12–18°C d) 40–45°C

11. Chickpea is a:
    a) Warm-season crop b) Cold-season crop
    c) Monsoon crop d) Tropical crop
12. Chickpea grows well at temperatures of:
    a) 10–15°C b) 20–30°C
    c) 30–40°C d) Above 40°C
13. Groundnut requires how much rainfall?
    a) 200–300 mm b) 500–1000 mm
    c) 1000–1500 mm d) >1500 mm
14. Optimum temperature range for sugarcane is:
    a) 10–15°C b) 18–22°C
    c) 26–32°C d) 40–45°C
15. Cotton requires how many days of frost-free weather?
    a) 50 b) 100
    c) 180–200 d) 300
16. Tomato is sensitive to:
    a) High humidity b) Temperature below 10°C
    c) Low sunlight d) Strong winds
17. Best temperature for tomato fruit setting is:
    a) 25–28°C b) 10–12°C
    c) 35–40°C d) >40°C
18. Potato grows well in which temperature range?
    a) 10–15°C b) 25–30°C
    c) 30–35°C d) >40°C
19. Mango is a:
    a) Temperate fruit b) Tropical fruit
    c) Alpine fruit d) Desert fruit
20. Citrus fruits require rainfall of:
    a) <200 mm b) 300–400 mm
    c) 750–1000 mm d) >2000 mm
21. Comfort zone temperature for dairy cattle is:
    a) 5–10°C b) 15–27°C
    c) 30–40°C d) >40°C

22. Heat stress in dairy cattle begins above:
    a) 20°C b) 25°C
    c) 30°C d) 15°C
23. Relative humidity above __% increases livestock stress:
    a) 40% b) 60%
    c) 70% d) 90%
24. Poultry birds are most comfortable at:
    a) 20–25°C b) 30–40°C
    c) 10–15°C d) >45°C
25. The temperature-humidity index (THI) is used to assess:
    a) Soil fertility b) Air quality
    c) Animal heat stress d) Milk quality
26. Goats are well adapted to
    a) Cold and humid regions b) Arid & semi-arid climates
    c) High rainfall areas d) Flood-prone zones
27. Sheep require an annual rainfall of about:
    a) 100–200 mm b) 200–400 mm
    c) 800–1000 mm d) >1500 mm
28. Broilers are best reared at temperatures of:
    a) 10–20°C b) 21–30°C
    c) >35°C d) <10°C
29. Buffaloes are more sensitive to:
    a) Cold stress b) Heat stress
    c) Wind d) Rain
30. High humidity leads to
    a) Decreased disease b) Increased feed conversion
    c) Reduced animal productivity d) Enhanced immunity
31. India is divided into how many agroclimatic zones (Planning Commission)?
    a) 10 b) 15
    c) 20 d) 15
32. In India, dryland agriculture is mainly practiced in:
    a) Coastal zones b) Humid tropics
    c) Semi-arid regions d) Wetlands

33. In arid zones, best suited crops are:
    a) Rice and maize
    b) Wheat and pulses
    c) Millets and oilseeds
    d) Sugarcane and banana
34. Eastern Himalayan region is suitable for:
    a) Wheat
    b) Tea
    c) Cotton
    d) Sorghum
35. Rainfed rice is majorly grown in:
    a) Western Rajasthan
    b) Peninsular India
    c) Punjab
    d) Haryana
36. Wheat cannot tolerate temperatures above:
    a) 20°C
    b) 30°C
    c) 10°C
    d) 5°C
37. Cold injury in paddy seedlings happens below:
    a) 10°C
    b) 20°C
    c) 25°C
    d) 30°C
38. Rainfed agriculture is possible with rainfall of:
    a) <200 mm
    b) 500–750 mm
    c) >2000 mm
    d) <100 mm
39. Sugarcane is highly sensitive to
    a) Frost
    b) Heat
    c) Light
    d) Salt
40. Optimum rainfall for groundnut is
    a) 200 mm
    b) 400 mm
    c) 500–1000 mm
    d) 1500 mm
41. Drought stress in crops leads to
    a) Increased leaf area
    b) Reduced stomatal conductance
    c) Higher photosynthesis
    d) Early flowering always
42. Heat stress during flowering in cereals causes:
    a) Enhanced grain filling
    b) Sterility and poor grain set
    c) Increased yield
    d) No effect
43. Frost damage mostly affects:
    a) Root crops
    b) Leafy vegetables & flowers
    c) Mature grains
    d) Livestock only

44. Excessive rainfall causes:
    a) Waterlogging and root rot b) Higher yields
    c) Increased nitrogen fixation d) Faster maturity
45. Salinity stress affects:
    a) Photosynthesis and water uptake b) Only root growth
    c) Fertilizer absorption only d) Pest resistance
46. Zebu cattle are more adapted to
    a) Cold climates b) Hot and humid climates
    c) Mountain regions d) Arctic regions
47. Shade provision reduces
    a) Feed intake b) Heat stress in animals
    c) Milk production d) Disease resistance
48. Water requirement in livestock increases during:
    a) Winter b) Summer
    c) Monsoon d) Autumn
49. Evaporative cooling in animals is primarily through:
    a) Sweating and panting b) Drinking water
    c) Increasing feed d) Huddling
50. Heat stress can cause
    a) Increased fertility
    b) Reduced milk yield and growth rate
    c) Decreased respiration
    d) Improved immunity

Answer Key

1	b	2	c	3	c	4	a	5	b	6	b	7	b
8	c	9	b	10	b	11	b	12	a	13	b	14	c
15	c	16	b	17	a	18	a	19	b	20	c	21	b
22	c	23	c	24	a	25	c	26	b	27	b	28	b
29	b	30	c	31	d	32	c	33	c	34	b	35	b
36	b	37	a	38	b	39	a	40	c	41	b	42	b
43	b	44	a	45	a	46	b	47	b	48	b	49	a
50	b												

Ch. 16. Weather forecasting

1. Weather forecasting is the prediction of:
   a) Soil fertility b) Past weather
   c) Atmospheric conditions d) Crop yield
2. The science that deals with weather forecasting is called:
   a) Climatology b) Geology
   c) Meteorology d) Agronomy
3. Which instrument is used to measure atmospheric pressure?
   a) Thermometer b) Barometer
   c) Hygrometer d) Anemometer
4. The organization responsible for weather forecasting in India is:
   a) NDMA b) IMD
   c) ICAR d) CSIR
5. Which satellite provides weather data to IMD?
   a) GSAT-1 b) INSAT-3D
   c) Aryabhata d) Mangalyaan
6. Nowcasting provides weather forecasts for a period of:
   a) Less than 12 hours b) 1–2 days
   c) 3–10 days d) Monthly
7. Short-range forecasts are valid for
   a) 1 hour b) 1–2 days
   c) 3–5 days d) 10–15 days
8. Medium-range forecasts are issued for
   a) Up to 3 days b) 4–10 days
   c) 10–20 days d) 1 month
9. Long-range forecasts are meant for
   a) Next 12 hours b) 1–2 days
   c) Weeks to seasons d) Daily weather
10. Seasonal forecasts are useful for
   a) Daily irrigation
   b) Sowing and cropping decisions
   c) Pesticide application
   d) Market planning

11. Numerical Weather Prediction (NWP) uses
    a) Farmers' observations   b) Empirical data only
    c) Mathematical models   d) Folk forecasting
12. Radar is primarily used to detect:
    a) Temperature   b) Humidity
    c) Rainfall and storms   d) Wind direction
13. Satellites provide which type of data for forecasting?
    a) Soil data
    b) Cloud movement and temperature
    c) Crop yield data
    d) Land ownership
14. Which model is most commonly used in global weather forecasting?
    a) ICAR Model   b) NWP Model
    c) GFS (Global Forecast System)   d) ATM Model
15. Remote sensing is helpful in:
    a) Soil testing   b) Fertilizer application
    c) Large-scale weather monitoring   d) Water treatment
16. Which forecast helps in deciding irrigation schedule?
    a) Seasonal forecast   b) Medium-range forecast
    c) Long-range forecast   d) Historical forecast
17. Pest and disease forecasting depends on:
    a) Soil fertility
    b) Weather variables like humidity, temperature
    c) Crop residues
    d) Market demand
18. Farmers can prevent losses by
    a) Ignoring weather data
    b) Following traditional calendars
    c) Using weather-based advisories
    d) Burning stubble
19. Weather forecasts help reduce
    a) Water harvesting   b) Input use efficiency
    c) Weather-related crop losses   d) Soil salinity

20. Weather forecast helps improve
    a) Rainfall uncertainty
    b) Market price
    c) Crop planning and risk management
    d) Pest resistance
21. Agromet Advisory Service is provided by:
    a) NDMA
    b) IMD
    c) FAO
    d) WTO
22. AAS provides:
    a) Daily market rates
    b) Crop insurance
    c) Weather-based agricultural advice
    d) Land ownership certificates
23. District Agromet Units (DAMUs) are established under:
    a) ICAR
    b) KVKs
    c) MoA
    d) ICRISAT
24. Which component is not included in AAS?
    a) Weather forecast
    b) Crop advisory
    c) Market support
    d) Livestock advisory
25. Farmers receive agromet advisories through:
    a) Newspapers only
    b) Television only
    c) Multi-channel communication including SMS, TV, and radio
    d) Email only
26. Which is not a weather parameter used in forecasting?
    a) Rainfall
    b) Temperature
    c) Wind speed
    d) Soil pH
27. Wind direction is measured using:
    a) Thermometer
    b) Hygrometer
    c) Wind vane
    d) Anemometer
28. Relative humidity is measured with a:
    a) Rain gauge
    b) Barometer
    c) Hygrometer
    d) Thermometer

29. Which of the following is a qualitative forecast?
    a) Rainfall in mm
    b) Cloud cover percentage
    c) Chance of rainfall (e.g., "likely")
    d) Wind speed in km/h
30. Pan evaporation data is used in:
    a) Forecasting soil health
    b) Crop damage estimation
    c) Irrigation scheduling
    d) Rain harvesting
31. Probability forecast expresses:
    a) Confidence in weather event
    b) Rain gauge accuracy
    c) Soil water level
    d) Forecast validity
32. Weather-based crop insurance is related to:
    a) MSP
    b) Pradhan Mantri Fasal Bima Yojana
    c) Swachh Bharat Mission
    d) NREGA
33. Forecasting helps in avoiding:
    a) Yield gains
    b) Timely sowing
    c) Weather-related risks
    d) Crop diversification
34. Rainfed farming benefits most from:
    a) Short-range forecast
    b) Seasonal forecast
    c) Market forecast
    d) Soil forecast
35. Which weather forecast helps with fertilizer application?
    a) Daily forecast
    b) Monthly forecast
    c) Long-range forecast
    d) Decadal forecast
36. One major challenge in weather forecasting is
    a) High costs
    b) Technological reliability
    c) Inaccuracy due to chaotic atmosphere
    d) Farmer illiteracy
37. Which is a limitation of long-range forecasts?
    a) Poor availability
    b) Lower precision
    c) Expensive instruments
    d) High temperature
38. Which factor limits forecast accuracy?
    a) Large datasets
    b) Too many satellites
    c) Unpredictable natural variability
    d) Farmer cooperation

39. Forecast accuracy reduces with
    a) Use of technology
    b) Larger data collection
    c) Longer time horizons
    d) Local observation
40. Effective forecasting needs integration of:
    a) Astrology and weather
    b) Climate models and ground data
    c) Soil testing
    d) Crop calendars only
41. Mobile apps for weather include
    a) mKisan, Meghdoot
    b) Krishi Vigyan
    c) eNAM
    d) Soil Health Card
42. Which is not a forecasting method?
    a) Synoptic
    b) Numerical
    c) Statistical
    d) Genetic
43. Which agency collaborates with IMD for AAS?
    a) WHO
    b) ISRO
    c) ICAR
    d) FAO
44. Forecasting monsoon onset is important for
    a) Sugarcane planting
    b) Kharif crop sowing
    c) Wheat irrigation
    d) Rabi harvest
45. Which sensor is used on satellites for weather data?
    a) GPS
    b) Radiometer
    c) X-ray sensor
    d) Altimeter
46. Use of AI in forecasting aims to:
    a) Remove satellites
    b) Improve forecast accuracy
    c) Increase rainfall
    d) Reduce temperature
47. Weather-based farming decisions lead to
    a) Increased risks
    b) Delayed farming
    c) Higher productivity and reduced losses
    d) More input cost
48. Customized weather advisories consider
    a) Market prices only
    b) Local weather & crop stages
    c) Universal guidelines
    d) Seed costs

49. Precision agriculture relies heavily on
    a) Historical records b) Weather forecasts & sensors
    c) Crop insurance d) Traditional methods
50. Main goal of weather forecasting in agriculture is:
    a) Marketing of inputs
    b) Crop insurance selling
    c) Risk management and sustainable farming
    d) Seed certification

Answer Key

1	c	2	c	3	b	4	b	5	b	6	a	7	b
8	b	9	c	10	b	11	b	12	c	13	c	14	c
15	c	16	b	17	b	18	c	19	c	20	c	21	b
22	c	23	b	24	c	25	c	26	d	27	c	28	c
29	c	30	c	31	a	32	b	33	c	34	b	35	a
36	c	37	b	38	c	39	c	40	b	41	a	42	d
43	c	44	b	45	b	46	b	47	c	48	b	49	b
50	c												

Ch. 17. Climate Change and Agriculture

1. Climate change refers to
    a) Daily weather changes
    b) Seasonal variations
    c) Long-term shifts in temperature and weather patterns
    d) Annual crop cycle
2. The main greenhouse gas responsible for global warming is:
    a) Nitrogen b) Carbon dioxide
    c) Oxygen d) Helium
3. Which gas is most potent in terms of global warming potential?
    a) Carbon dioxide b) Methane
    c) Oxygen d) Nitrogen
4. Global warming is mainly caused by:
    a) Solar flares b) Greenhouse gas emissions
    c) Plate tectonics d) Volcanoes

5. Which sector contributes significantly to greenhouse gas emissions?
   a) Banking b) Agriculture
   c) Education d) IT
6. The primary source of methane in agriculture is:
   a) Fertilizer application
   b) Livestock enteric fermentation
   c) Soil erosion
   d) Wind erosion
7. Nitrous oxide is released from agricultural fields mainly due to:
   a) Burning residues b) Use of nitrogen fertilizers
   c) Pesticide spraying d) Water irrigation
8. Which gas has the highest global warming potential (GWP)?
   a) CO_2 b) N_2O
   c) CH_3 d) H_2O
9. Livestock farming contributes to climate change through
   a) Methane emission b) Soil compaction
   c) Biodiversity d) Groundwater recharge
10. Which practice helps reduce methane from paddy fields?
    a) Continuous flooding b) Alternate wetting & drying
    c) Late transplanting d) Using more fertilizer
11. Climate change affects agriculture by
    a) Improving soil fertility
    b) Increasing productivity everywhere
    c) Altering rainfall and temperature patterns
    d) Increasing water availability
12. Heat stress due to climate change can reduce
    a) Fertilizer efficiency b) Crop pollination and yield
    c) Labour availability d) Soil erosion
13. An increase in CO_2 concentration generally leads to
    a) Higher stomatal opening
    b) Increased photosynthesis in C3 crops
    c) Less water use efficiency
    d) Reduction in plant growth

14. Crop most sensitive to temperature rise is
    a) Sorghum b) Rice
    c) Wheat d) Maize
15. Climate change may lead to more frequent
    a) Cloudy days b) Droughts and floods
    c) Monsoons d) Fertility increase
16. Adaptation to climate change in agriculture includes
    a) Carbon trading
    b) Developing drought-tolerant varieties
    c) Buying machinery
    d) Land use change
17. Mitigation means
    a) Adjusting to new conditions
    b) Reducing causes of climate change
    c) Increasing crop diversity
    d) Building irrigation canals
18. Carbon sequestration refers to
    a) Emitting CO_2
    b) Capturing and storing atmospheric CO_2
    c) Growing tall plants
    d) Measuring carbon in soil
19. Agroforestry helps mitigate climate change by
    a) Increasing evapotranspiration
    b) Reducing biodiversity
    c) Absorbing atmospheric carbon dioxide
    d) Releasing greenhouse gases
20. Which farming practice is climate-smart?
    a) Slash-and-burn b) Zero tillage
    c) Mono-cropping d) Overgrazing
21. India's National Action Plan on Climate Change (NAPCC) has how many missions?
    a) 3 b) 5
    c) 8 d) 10

22. National Mission for Sustainable Agriculture (NMSA) is part of:
    a) Green Revolution
    b) NAPCC
    c) PMFBY
    d) Rashtriya Krishi Vikas Yojana
23. UNFCCC stands for
    a) United Nations Framework Convention on Climate Change
    b) United Nations Food and Climate Committee
    c) Union National Federation of Climate Change
    d) None of these
24. The Kyoto Protocol aimed to reduce
    a) Soil erosion
    b) Global trade
    c) Greenhouse gas emissions
    d) Ocean salinity
25. Paris Agreement's key goal is to limit global warming to:
    a) 1.5–2°C above pre-industrial levels
    b) 3°C above industrial levels
    c) 2.5°C above average
    d) Below 5°C always
26. Which type of crop is better suited for dry conditions?
    a) Paddy
    b) Sugarcane
    c) Sorghum
    d) Jute
27. Mulching helps in
    a) Increasing evaporation
    b) Reducing soil moisture loss
    c) Enhancing soil salinity
    d) Water logging
28. SRI (System of Rice Intensification) helps in
    a) Using more water
    b) Increasing methane emissions
    c) Reducing water use and emissions
    d) Growing only basmati rice
29. Crop diversification is a strategy for
    a) Disease increase
    b) Mitigation of climate change risks
    c) Less productivity
    d) Monoculture

30. Drought-tolerant crops include
    a) Maize
    b) Groundnut
    c) Pearl millet
    d) Sugarcane
31. GCMs in climate science stands for
    a) General Climate Measures
    b) Global Crop Models
    c) General Circulation Models
    d) Ground Control Measures
32. Agrometeorology helps in
    a) Assessing soil nutrition
    b) Predicting crop pests
    c) Understanding weather impact on crops
    d) Testing new machines
33. Climate scenarios are used to
    a) Evaluate new seeds
    b) Predict market prices
    c) Assess future climate impacts
    d) Measure yield directly
34. Which is not a climate model output?
    a) Rainfall patterns
    b) CO_2 levels
    c) Pest counts
    d) Temperature trends
35. One limitation of climate models is
    a) Too accurate
    b) Fixed assumptions
    c) No global application
    d) No scientific basis
36. Climate change increases risk of
    a) Food security
    b) Water scarcity
    c) Better infrastructure
    d) Market stability
37. Farmers in rainfed areas are more vulnerable due to:
    a) Better irrigation
    b) Lack of climate resilience
    c) Fertile soils
    d) Access to loans
38. Extreme weather events affect
    a) Soil pH
    b) Harvest and storage
    c) Remotesensing
    d) Seed certification
39. Migration of rural populations is linked to
    a) Urban growth only
    b) Climate-induced livelihood losses
    c) Industrialization only
    d) Better transport

40. Gender is important in climate discussions because:
    a) Women are less educated
    b) Men are less affected
    c) Women are disproportionately impacted
    d) Only men are farmers
41. Carbon footprint in farming refers to:
    a) Use of machinery
    b) Total GHG emissions from agricultural activities
    c) Soil erosion
    d) Biodiversity loss
42. Renewable energy in agriculture includes
    a) Diesel pumps
    b) Solar dryers
    c) Electric tractors
    d) Tube wells
43. Energy-efficient practices help in
    a) Increasing emissions
    b) Climate change mitigation
    c) Decreasing yield
    d) Polluting water
44. Which source is not renewable?
    a) Wind
    b) Coal
    c) Solar
    d) Biogas
45. Energy audit in farming helps in
    a) Financial planning
    b) Emission estimation and efficiency improvement
    c) Water conservation
    d) Soil testing
46. Climate-smart agriculture aims for
    a) Higher input use
    b) Resilience, mitigation, and productivity
    c) Monocropping
    d) Export growth
47. Which of the following is NOT a climate-smart practice?
    a) Integrated pest management
    b) Overuse of fertilizers
    c) Crop rotation
    d) Rainwater harvesting

48. Which term refers to farming adapted to changing climate?
    a) Climate-resilient agriculture b) Irrigated farming
    c) Organic farming d) Commercial farming
49. Digital tools in climate-smart farming help in
    a) Manual sowing b) Real-time weather advisories
    c) Reducing technology use d) Pest breeding
50. Which Indian mission promotes climate-resilient agriculture?
    a) National Food Security Mission b) NMSA
    c) NRLM d) MGNREGA

Answer Key

1	c	2	b	3	b	4	b	5	b	6	b	7	b
8	b	9	a	10	b	11	c	12	b	13	b	14	c
15	b	16	b	17	b	18	b	19	c	20	b	21	c
22	b	23	a	24	c	25	a	26	c	27	b	28	c
29	b	30	c	31	c	32	c	33	c	34	c	35	b
36	b	37	b	38	b	39	b	40	c	41	b	42	b
43	b	44	b	45	b	46	b	47	b	48	a	49	b
50	b												

Ch. 18. Weather and Agriculture Relationship

1. Weather refers to
    a) Long-term atmospheric conditions
    b) Daily atmospheric conditions
    c) Average temperature only
    d) Seasonal crop yield
2. Climate refers to:
    a) Temporary weather variations
    b) Long-term average of weather conditions
    c) Soil fertility
    d) Rainfall on a single day
3. Which of the following is not an element of weather?
    a) Temperature b) Humidity
    c) Soil pH d) Wind speed

4. Agrometeorology is the science of:
   a) Soil testing
   b) Relationship between weather and crop production
   c) Animal health
   d) Insecticide formulation
5. Which weather element most directly affects crop evapotranspiration?
   a) Soil color
   b) Wind speed
   c) Humidity
   d) All of the above
6. The crop stage most sensitive to adverse weather is
   a) Vegetative
   b) Germination
   c) Flowering
   d) Ripening
7. Low night temperatures in wheat during grain filling cause
   a) Poor germination
   b) Better grain yield
   c) Leaf curling
   d) Sterility
8. Which of the following crops is most sensitive to frost?
   a) Sorghum
   b) Tomato
   c) Pearl millet
   d) Cotton
9. Rainfall deficiency during reproductive stage of crops causes
   a) Enhanced seed size
   b) Better root development
   c) Yield reduction
   d) Delayed maturity
10. Excessive humidity is favorable for
    a) Drought-resistant crops
    b) Disease development
    c) Improved photosynthesis
    d) Early flowering
11. Crops that grow in high humidity are called
    a) Xerophytes
    b) Hydrophytes
    c) Mesophytes
    d) Hygrophytes
12. Optimum temperature for wheat germination is
    a) 10–15°C
    b) 20–25°C
    c) 30–35°C
    d) 5–10°C
13. Photosynthesis increases with
    a) Increasing light and temperature (within limits)
    b) Decreasing CO2
    c) High wind speed
    d) High soil moisture alone

14. A weather parameter essential for calculating growing degree days is:
    a) Humidity
    b) Soil moisture
    c) Temperature
    d) Rainfall
15. Maize is sensitive to water stress during
    a) Germination
    b) Vegetative stage
    c) Silking stage
    d) Maturity
16. Heatwaves are particularly damaging to
    a) Rice flowering
    b) Wheat grain filling
    c) Maize germination
    d) Sugarcane tillering
17. Hailstorms cause major damage to
    a) Root crops
    b) Fruit and vegetable crops
    c) Millets
    d) Legumes
18. Drought conditions mainly affect
    a) Root growth
    b) Water availability & yield
    c) Soil structure
    d) Solar radiation
19. Frost damage is primarily due to
    a) High humidity
    b) Low temperature below freezing
    c) High wind speed
    d) Heavy rainfall
20. Stormy weather affects agriculture by:
    a) Enhancing pollination
    b) Improving soil fertility
    c) Causing lodging in crops
    d) Promoting flowering
21. Land preparation is best done during:
    a) High rainfall period
    b) Dry weather
    c) Post-harvest stage
    d) Cold waves
22. Delayed sowing due to unexpected rainfall affects:
    a) Soil structure
    b) Weed control
    c) Crop duration and yield
    d) Harvest quality
23. Harvesting in rainy weather causes:
    a) Easy threshing
    b) Crop spoilage and fungal attack
    c) High seed viability
    d) Reduced storage losses

24. Dry weather is ideal for:
    a) Sowing
    b) Harvesting and drying
    c) Irrigation
    d) Transplanting
25. Weather affects pest dynamics by
    a) Killing them directly
    b) Modifying their life cycle and survival
    c) Altering seed variety
    d) Reducing fertilizer need
26. High ambient temperatures in livestock cause
    a) Increased appetite
    b) Heat stress and reduced productivity
    c) Higher milk yield
    d) Lower respiration rate
27. Cold stress in poultry causes
    a) Egg production increase
    b) Reduced feed intake
    c) High mortality
    d) Poor feathering
28. Fish production is directly affected by
    a) Air pollution
    b) Wind direction
    c) Water temperature & rainfall
    d) Cloud cover
29. Extreme rainfall events affect livestock by
    a) Increasing milk production
    b) Causing waterlogging and diseases
    c) Enhancing digestion
    d) Reducing fodder quality
30. Heat waves affect dairy animals by
    a) Increasing conception rate
    b) Enhancing feed quality
    c) Reducing milk yield
    d) No impact
31. Weather data helps farmers to
    a) Choose fertilizers
    b) Plan sowing and irrigation
    c) Increase subsidies
    d) Decide crop rotation
32. Which forecast is most useful for crop planning?
    a) Nowcast
    b) Medium-range forecast
    c) Seasonal forecast
    d) Long-term prediction

33. Dry spells after sowing affect
    a) Seed germination
    b) Pest infestation
    c) Leaf shedding
    d) Weed suppression
34. Which crop can tolerate high rainfall?
    a) Groundnut
    b) Maize
    c) Rice
    d) Cotton
35. Weather aberrations are
    a) Predictable temperature changes
    b) Deviations from normal weather patterns
    c) Daily fluctuations
    d) Expected seasonal shifts
36. Irrigation scheduling is best done based on
    a) Soil color
    b) Temperature alone
    c) Evapotranspiration & rainfall
    d) Pest infestation
37. Weeding operations are best done in
    a) Flooded conditions
    b) Windy weather
    c) Moist but not wet soils
    d) Frozen soils
38. Excess rain after fertilizer application causes
    a) Nutrient enhancement
    b) Leaching losses
    c) Improved uptake
    d) Higher chlorophyll
39. Crop yield forecasting depends heavily on:
    a) Weather and crop models
    b) Fertilizer type
    c) Pest load
    d) Seed size
40. Cloudy weather generally results in:
    a) Higher radiation
    b) Reduced photosynthesis
    c) Greater transpiration
    d) Early maturity
41. Contingency crop planning is required for:
    a) Mechanization
    b) Weather aberrations
    c) Increasing subsidies
    d) Extending harvest
42. Crop insurance is influenced by:
    a) Soil type
    b) Weather risks
    c) Irrigation facility
    d) Market prices

43. Weather-based advisory helps in
   a) Policy decisions only
   b) On-farm operational planning
   c) Machinery scheduling
   d) Market analysis
44. Drought-prone areas require
   a) Heavy irrigation
   b) Short-duration, drought-tolerant crops
   c) Transplanting rice
   d) High-input farming
45. Real-time weather data helps in
   a) Historical analysis only
   b) Immediate decision making on farm operations
   c) Reducing soil testing
   d) None of the above
46. IMD provides weather forecasts through
   a) Krishi Vigyan Kendras
   b) SMS and mobile apps
   c) Banks
   d) Panchayats only
47. Which mobile app provides agromet advisories in India?
   a) Arogya Setu
   b) Meghdoot
   c) Kisan Seva
   d) Pusa Krishi
48. Which of these is a benefit of weather forecasting in agriculture?
   a) Delay in sowing
   b) Efficient input use
   c) Increased manual labor
   d) Lower seed quality
49. Farmers receive weather advisories through:
   a) Radio
   b) SMS
   c) TV
   d) All of the above
50. Which is an agrometeorological service?
   a) Soil survey
   b) Weather-based crop advisories
   c) Credit schemes
   d) Seed certification

Answer Key

1	b	2	b	3	c	4	b	5	d	6	c	7	b
8	b	9	c	10	b	11	d	12	a	13	a	14	c
15	c	16	b	17	b	18	b	19	b	20	c	21	b
22	c	23	b	24	b	25	b	26	b	27	c	28	c
29	b	30	c	31	b	32	b	33	a	34	c	35	b
36	c	37	c	38	b	39	a	40	b	41	b	42	b
43	b	44	b	45	b	46	b	47	b	48	b	49	d
50	b												

Ch. 19. Impact of Climate Change and Adaptation Strategies in Horticulture

1. Climate change primarily affects horticulture through:
   a) Soil pH
   b) Changes in temperature and rainfall patterns
   c) Increase in seed production
   d) Reduction in pests
2. One major effect of climate change on fruit crops is:
   a) Enhanced flowering time
   b) Shift in flowering and fruiting seasons
   c) Constant yield
   d) Increased resistance to pests
3. Global warming can cause:
   a) Decreased evapotranspiration
   b) Increased incidence of heat stress in horticultural crops
   c) Lower soil temperature
   d) Longer dormancy periods
4. Increased atmospheric CO_2 concentration affects horticultural crops by:
   a) Decreasing photosynthesis
   b) Increasing photosynthesis and water use efficiency
   c) Reducing biomass
   d) Causing plant death

5. Extreme weather events under climate change can:
   a) Improve crop resilience
   b) Cause crop damage and yield loss
   c) Increase market price stability
   d) Eliminate pests
6. High temperature stress during flowering in horticultural crops causes:
   a) Increased fruit set b) Flower drop & poor fruit set
   c) Early harvest d) Enhanced fruit quality
7. Drought conditions due to climate change reduce
   a) Fruit size and quality b) Photosynthesis rate only
   c) Pest incidence d) Root growth
8. Heat waves can lead to
   a) Increased fruit sugar content
   b) Sunburn damage on fruits and leaves
   c) Delayed maturity
   d) Higher fruit acidity
9. Waterlogging from heavy rainfall results in:
   a) Increased root respiration
   b) Root hypoxia and disease susceptibility
   c) Better nutrient uptake
   d) Increased flowering
10. Irrigation scheduling is critical under climate change to:
    a) Prevent soil erosion
    b) Maintain optimum soil moisture
    c) Increase pest resistance
    d) Promote flowering
11. Rising temperatures generally cause
    a) Decreased pest population
    b) Increased pest and disease outbreaks
    c) Elimination of diseases
    d) Increased natural enemies only

12. Humidity changes due to climate change can:
    a) Decrease fungal diseases
    b) Increase fungal and bacterial infections
    c) Prevent viral diseases
    d) Reduce pest infestation
13. Increased CO_2 affects pest dynamics by:
    a) Increasing pest reproduction rates
    b) Killing pests
    c) Making crops toxic
    d) Increasing pesticide efficiency
14. Warmer winters lead to
    a) Reduced pest survival
    b) Higher overwintering survival of pests
    c) Decreased insect outbreaks
    d) Pest migration to poles
15. Climate change adaptation strategies for pest control include:
    a) Ignoring pest outbreaks
    b) Integrated Pest Management (IPM)
    c) Using chemical pesticides only
    d) Monocropping
16. Climate change can cause
    a) Shift in flowering and fruiting time
    b) Uniform crop cycles
    c) Decrease in crop biodiversity
    d) Constant yield
17. Elevated CO_2 levels can increase:
    a) Yield potential of some horticultural crops
    b) Crop vulnerability to frost
    c) Soil salinity
    d) Pest incidence only
18. Heat stress can reduce fruit yield by:
    a) Reducing fruit set and quality
    b) Increasing leaf growth
    c) Extending growing season
    d) Increasing pollination

19. Fruit quality under climate change is often:
    a) Improved in all crops
    b) Reduced due to stress conditions
    c) Unaffected
    d) More uniform
20. Phenological shifts due to climate change affect:
    a) Water use efficiency
    b) Pest cycles and pollination timing
    c) Soil fertility
    d) Crop color only
21. Efficient water use under climate change can be achieved by:
    a) Flood irrigation
    b) Drip and micro-irrigation
    c) Increasing irrigation frequency without monitoring
    d) Rainfed only
22. Mulching helps adaptation by
    a) Increasing soil temperature
    b) Reducing evaporation and conserving soil moisture
    c) Increasing weed growth
    d) Increasing runoff
23. Rainwater harvesting benefits horticulture by
    a) Reducing groundwater recharge
    b) Providing supplemental irrigation water
    c) Increasing soil erosion
    d) Enhancing pest survival
24. Scheduling irrigation based on crop growth stage:
    a) Wastes water
    b) Saves water and improves yield
    c) Reduces soil fertility
    d) Increases salinity
25. Soil moisture sensors help:
    a) Monitor irrigation needs precisely b) Increase evaporation
    c) Detect pests d) Change crop variety

26. Selection of climate-resilient varieties means:
    a) Choosing drought and heat tolerant genotypes
    b) Using traditional varieties only
    c) Ignoring climate data
    d) Planting monocultures
27. Adjusting planting dates can
    a) Avoid peak stress periods
    b) Reduce yields
    c) Increase pest outbreaks
    d) Reduce soil fertility
28. Use of shade nets in horticulture
    a) Increases heat stress
    b) Protects crops from excessive heat and UV radiation
    c) Reduces photosynthesis
    d) Enhances wind damage
29. Crop diversification helps adaptation by
    a) Increasing income risk
    b) Reducing vulnerability to climate stress
    c) Increasing pest outbreaks
    d) Reducing biodiversity
30. Soil health management under climate change includes:
    a) Avoiding organic amendments
    b) Using organic mulches and biofertilizers
    c) Using chemical fertilizers only
    d) Removing soil cover
31. Climate change affects post-harvest by
    a) Increasing shelf life
    b) Increasing post-harvest losses due to higher temperatures
    c) Reducing transportation costs
    d) Improving packaging
32. Cold storage and controlled atmosphere storage help by:
    a) Accelerating spoilage
    b) Reducing post-harvest losses
    c) Increasing water loss
    d) Promoting diseases

33. Value addition and processing reduce
    a) Market price
    b) Post-harvest losses and improve income
    c) Crop quality
    d) Farmer income
34. Developing early maturing varieties helps
    a) Avoid late-season drought and heat stress
    b) Increase water use
    c) Increase pests
    d) Extend cropping season only
35. Crop insurance schemes help farmers by
    a) Ignoring climate risks
    b) Providing financial support against climate-induced losses
    c) Increasing input cost
    d) Limiting crop choice
36. Climate-smart agriculture promotes
    a) Increased use of pesticides
    b) Sustainability, productivity, and resilience
    c) Monocropping
    d) Deforestation
37. Farmer awareness and training on climate change
    a) Has no impact
    b) Improves adoption of adaptation practices
    c) Only benefits scientists
    d) Reduces productivity
38. Access to weather forecasts helps farmers
    a) Ignore climate variability
    b) Plan field activities and reduce risks
    c) Increase crop failures
    d) Reduce irrigation

39. Government subsidies for micro-irrigation
    a) Discourage water saving
    b) Promote water-efficient farming
    c) Increase water wastage
    d) Are irrelevant
40. Community-based adaptation includes:
    a) Individual farming only
    b) Collective resource management and knowledge sharing
    c) Ignoring local conditions
    d) Limiting technology use
41. Use of remote sensing and GIS helps
    a) Monitor crop stress and manage resources efficiently
    b) Increase input costs only
    c) Reduce data availability
    d) Ignore weather changes
42. Biotechnological approaches in horticulture aim to:
    a) Develop stress tolerant varieties b) Increase pesticide use
    c) Reduce genetic diversity d) Delay maturity
43. Use of protected cultivation like polyhouses:
    a) Protects crops from climate extremes
    b) Increases water loss
    c) Reduces yields
    d) Increases pest attacks
44. Precision farming helps adaptation by
    a) Applying inputs uniformly
    b) Site-specific management based on climate and soil data
    c) Increasing waste
    d) Ignoring weather data
45. Early warning systems for pests and weather help:
    a) Increase crop damage
    b) Reduce risks by timely action
    c) Delay harvest
    d) Increase input costs

46. Diversifying income sources reduces:
    a) Climate vulnerability
    b) Farmer income
    c) Market access
    d) Food security
47. Access to credit and insurance enables
    a) Risk taking and adaptation investments
    b) Crop failure
    c) Increased input cost
    d) Reduced productivity
48. Farmers' participation in decision-making:
    a) Decreases adaptation success
    b) Enhances relevance and uptake of adaptation strategies
    c) Limits innovation
    d) Reduces extension services
49. Gender mainstreaming in adaptation:
    a) Is irrelevant
    b) Ensures inclusive climate resilience
    c) Restricts women's role
    d) Ignores women farmers
50. Market linkages improve adaptation by:
    a) Increasing price volatility
    b) Providing stable income and access to inputs
    c) Reducing crop diversity
    d) Increasing losses

Answer Key

1	b	2	b	3	b	4	b	5	b	6	b	7	a
8	b	9	b	10	b	11	b	12	b	13	a	14	b
15	b	16	a	17	a	18	a	19	b	20	b	21	b
22	b	23	b	24	b	25	a	26	a	27	a	28	b
29	b	30	b	31	b	32	b	33	b	34	a	35	b
36	b	37	b	38	b	39	b	40	b	41	a	42	a
43	a	44	b	45	b	46	a	47	a	48	b	49	b
50	b												

Ch. 20. Impact of Climate Change on Sericulture

1. Climate change affects sericulture primarily by altering:
   a) Soil fertility
   b) Temperature and humidity regimes
   c) Silkworm genetics
   d) Mulberry leaf color
2. Optimal temperature for mulberry leaf growth is around:
   a) 10–15°C b) 24–28°C
   c) 30–35°C d) 40–45°C
3. Increased temperature due to climate change can:
   a) Enhance mulberry growth indefinitely
   b) Cause heat stress reducing leaf yield
   c) Decrease silkworm growth
   d) Increase cocoon weight
4. High temperature and low humidity adversely affect:
   a) Mulberry leaf quality b) Silkworm breeding only
   c) Cocoon shell thickness d) Soil pH
5. Climate change causes shift in:
   a) Silkworm crop cycles and breeding seasons
   b) Silkworm genetic makeup
   c) Soil texture
   d) Silk market prices
6. Mulberry is sensitive to:
   a) High temperature & drought b) Cold only
   c) Waterlogging only d) Soil salinity only
7. Drought stress reduces:
   a) Leaf area & moisture content b) Leaf thickness only
   c) Leaf color d) Stem length only
8. Excessive rainfall leads to:
   a) Root rot and fungal diseases b) Increased leaf yield
   c) Improved leaf quality d) Reduced pest incidence

9. Mulberry leaves grown under heat stress show:
   a) Higher protein content
   b) Reduced nutritional quality for silkworms
   c) No change
   d) Increased fiber content
10. Mulberry yield is closely linked to:
   a) Soil pH only
   b) Temperature and rainfall patterns
   c) Pest control only
   d) Fertilizer type only
11. Silkworms are most sensitive to temperature during:
   a) Egg incubation and larval growth
   b) Cocoon spinning only
   c) Adult moth stage
   d) None of the above
12. Ideal temperature for silkworm larval rearing is:
   a) 15–20°C   b) 23–28°C
   c) 30–35°C   d) >35°C
13. High temperature (>30°C) causes:
   a) Faster larval development but poor cocoon quality
   b) Larger cocoons
   c) Longer larval period
   d) Increased silk quality
14. Low humidity (<70%) during silkworm rearing causes:
   a) Better cocoon shell formation
   b) Larval dehydration and mortality
   c) Faster larval growth
   d) Increased cocoon weight
15. Fluctuations in temperature cause
   a) Stable silkworm growth
   b) Stress leading to lower silk yield
   c) Improved cocoon quality
   d) None of the above

16. Climate change can increase incidence of:
    a) Viral and fungal diseases in silkworms
    b) Only bacterial diseases
    c) No diseases
    d) Only pest infestations
17. High humidity favors:
    a) Bacterial diseases in silkworms
    b) Fungal infections such as grasserie
    c) Reduced disease incidence
    d) Increased cocoon production
18. Extreme temperatures can
    a) Reduce silkworm immunity, increasing disease susceptibility
    b) Increase resistance to disease
    c) Have no effect on diseases
    d) Kill all pathogens
19. Grasserie disease is aggravated by:
    a) High temperature & humidity
    b) Low temperature only
    c) Dry conditions only
    d) None of the above
20. Disease management under climate stress requires:
    a) Frequent chemical sprays only
    b) Integrated disease management including environment control
    c) Ignoring weather changes
    d) Increasing silkworm density
21. Drought-tolerant mulberry varieties help by
    a) Increasing water requirement
    b) Reducing crop failure under water stress
    c) Increasing pest incidence
    d) Reducing leaf quality
22. Mulberry mulching:
    a) Increases soil moisture loss
    b) Conserves soil moisture and reduces temperature stress
    c) Has no effect on climate stress
    d) Increases weed growth

23. Use of drip irrigation in mulberry cultivation
    a) Saves water and improves leaf yield
    b) Wastes water
    c) Increases waterlogging
    d) Reduces leaf quality
24. Changing planting dates in mulberry
    a) Avoids peak heat and drought periods
    b) Increases pest attack
    c) Reduces leaf quality
    d) None of the above
25. Soil organic matter management helps
    a) Improve water retention and nutrient availability
    b) Increase soil erosion
    c) Reduce fertility
    d) Increase pest incidence
26. Climate-controlled rearing houses help
    a) Maintain optimum temperature and humidity
    b) Increase disease incidence
    c) Increase water consumption
    d) Reduce cocoon quality
27. Use of improved silkworm breeds
    a) Enhances tolerance to temperature fluctuations
    b) Decreases yield
    c) Increases disease susceptibility
    d) None of the above
28. Regular monitoring of microclimate in rearing houses:
    a) Helps early detection of stress conditions
    b) Is unnecessary
    c) Increases costs only
    d) Reduces yield

29. Improved disease management under climate stress:
    a) Uses bio-pesticides and sanitation
    b) Relies on chemicals only
    c) Ignores environmental control
    d) None of the above
30. Adjusting feeding schedules helps to:
    a) Reduce heat stress in silkworms
    b) Increase labor cost only
    c) Have no effect on silkworm health
    d) Reduce silk quality
31. Climate change impacts on sericulture can lead to:
    a) Increased farmer income
    b) Reduced cocoon production and farmer livelihoods
    c) Stable markets
    d) Increased exports
32. Farmers' awareness of climate risks in sericulture:
    a) Is not important
    b) Helps adoption of adaptation practices
    c) Increases costs only
    d) None of the above
33. Access to credit for climate-resilient technologies:
    a) Facilitates better adaptation
    b) Has no impact
    c) Reduces adoption
    d) Increases vulnerability
34. Diversification in sericulture includes:
    a) Mono cropping only
    b) Intercropping mulberry with other crops
    c) Ignoring climate data
    d) Increasing silkworm density only
35. Government policies supporting climate-smart sericulture:
    a) Increase vulnerability
    b) Promote resilience and sustainability
    c) Limit farmer options
    d) Are irrelevant

36. Development of heat-tolerant silkworm breeds:
    a) Is essential for climate change adaptation
    b) Reduces silk quality
    c) Increases disease incidence
    d) Is not possible
37. Use of mulberry varieties tolerant to abiotic stresses:
    a) Enhances sustainability b) Has no effect on yield
    c) Increases input cost only d) Is irrelevant
38. Remote sensing in sericulture helps to:
    a) Monitor mulberry crop health and water stress
    b) Increase labor cost only
    c) Reduce leaf quality
    d) Have no practical use
39. Biotechnological approaches include:
    a) Genetic improvement of silkworm and mulberry
    b) Only chemical use
    c) Ignoring environmental factors
    d) None of the above
40. Protected cultivation for mulberry:
    a) Minimizes climate stress effects b) Increases water loss
    c) Reduces leaf yield d) Increases pest attack
41. Climate change may alter pest species composition in sericulture by:
    a) Reducing pest diversity
    b) Shifting pest populations and emergence of new pests
    c) Eliminating all pests
    d) None of the above
42. Maintaining biodiversity in mulberry plantations:
    a) Helps ecological balance and pest control
    b) Increases disease risk
    c) Reduces yields
    d) Has no effect
43. Soil erosion due to extreme rainfall can:
    a) Reduce mulberry productivity b) Improve soil quality
    c) Have no effect d) Increase leaf size

44. Climate change can lead to:
    a) Increased soil salinity in some regions affecting mulberry growth
    b) Decreased soil temperature only
    c) Soil fertility increase
    d) None of the above
45. Conservation of water and soil in mulberry gardens:
    a) Is crucial for climate resilience
    b) Has no benefit
    c) Increases production cost only
    d) None of the above
46. Integrated management for climate-resilient sericulture includes:
    a) Crop, pest, and water management combined
    b) Chemical use only
    c) Ignoring climate factors
    d) No changes
47. Climate adaptation in sericulture requires:
    a) Only farmer knowledge
    b) Coordination among researchers, extension, and farmers
    c) No research
    d) Increased costs only
48. Monitoring climatic parameters in sericulture helps
    a) Ignore weather changes
    b) Take timely corrective measures
    c) Increase risks
    d) Reduce yields
49. Climate change mitigation in sericulture involves:
    a) Reducing greenhouse gas emissions and carbon sequestration
    b) Increasing chemical fertilizers only
    c) Expanding cultivation into fragile areas
    d) Ignoring environmental impact
50. Future sericulture sustainability depends on:
    a) Technological innovations and adaptive management
    b) Traditional practices only
    c) Ignoring climate data
    d) Increasing monocropping

Answer Key

1	b	2	b	3	b	4	a	5	a	6	a	7	a
8	a	9	b	10	b	11	a	12	b	13	a	14	b
15	b	16	a	17	b	18	a	19	a	20	b	21	b
22	b	23	a	24	a	25	a	26	a	27	a	28	a
29	a	30	a	31	b	32	b	33	a	34	b	35	b
36	a	37	a	38	a	39	a	40	a	41	a	42	b
43	a	44	a	45	a	46	a	47	b	48	b	49	a
50	a												

Ch. 21. Agricultural Meteorology: An Essential Associate for Food and Nutrition Security

1. Agricultural Meteorology primarily deals with:
   a) Soil fertility
   b) Weather impacts on crops and livestock
   c) Farm mechanization
   d) Market trends
2. The main aim of agricultural meteorology is to:
   a) Improve rainfall patterns
   b) Study diseases in livestock
   c) Enhance food and nutrition security through weather-informed practices
   d) Increase landholdings
3. Which of the following is a key atmospheric element in agrometeorology?
   a) Soil color b) Wind direction
   c) Seed variety d) Fertilizer type
4. Which organization provides agromet advisories in India?
   a) NABARD b) FSSAI
   c) IMD (India Meteorological Dept) d) WTO
6. Which stage of crop is most vulnerable to weather variability?
   a) Vegetative b) Flowering
   c) Maturity d) Storage
7. Prolonged drought during the cropping season can lead to:
   a) Reduced pest attack b) Higher yield
   c) Crop failure d) Delayed flowering

8. Solar radiation directly influences:
   a) Leaf senescence b) Crop photosynthesis
   c) Soil acidity d) Harvest index
9. Evapotranspiration is the sum of evaporation and:
   a) Infiltration b) Rainfall
   c) Transpiration d) Irrigation
10. Which crop is more sensitive to low temperature stress?
   a) Pearl millet b) Wheat
   c) Sorghum d) Sugarcane
11. Food security means
   a) Adequate availability of food only
   b) Access to junk food
   c) Access to sufficient, safe and nutritious food
   d) Free food supply
12. Malnutrition can result from
   a) Balanced diet
   b) Weather-driven food production losses
   c) Reduced rainfall variability
   d) Pest control
13. Weather-induced crop failure impacts nutrition by:
   a) Increasing calorie consumption
   b) Enhancing food diversity
   c) Reducing food availability and quality
   d) Increasing soil fertility
14. Climate-resilient crops are promoted to
   a) Increase dependence on imports
   b) Maintain food supply under changing climates
   c) Reduce productivity
   d) Make farming more expensive
15. Food insecurity is highest in
   a) Coastal cities
   b) Mountain towns
   c) Drought-prone and flood-prone rural areas
   d) Industrial zones

16. Agricultural meteorology plays a major role in combating:
    a) Waterlogging
    b) Climate change impacts on agriculture
    c) Market failures
    d) Land reforms
17. Global warming primarily affects
    a) Soil fertility
    b) Weather extremes
    c) Machinery usage
    d) Seed prices
18. Increased CO_2 levels may lead to
    a) Better nutrition in all crops
    b) Reduced crop yields under stress
    c) Lower photosynthesis
    d) Immediate rainfall
19. Mitigation of climate change through agriculture includes:
    a) Burning residues
    b) Precision farming and reduced emissions
    c) Flood irrigation
    d) Excess fertilizer use
20. Which of the following is an adaptation strategy?
    a) Switching to longer duration varieties
    b) Use of weather-based advisories
    c) Ignoring weather data
    d) Monocropping
21. Erratic weather can cause:
    a) Stable food production
    b) Global food price volatility
    c) Soil formation
    d) Water desalination
22. The main driver of global hunger is:
    a) Food storage issues
    b) Political conflict only
    c) Climate extremes & poor access
    d) Internet issues
23. Which continent is most vulnerable to food insecurity from weather change?
    a) Europe
    b) North America
    c) Africa
    d) Australia

24. Extreme rainfall events can cause
    a) Improved soil aeration
    b) Waterlogging & crop damage
    c) Higher nitrogen use
    d) Better storage conditions
25. Resilient agriculture aims to
    a) Maintain productivity despite weather challenges
    b) Depend on old methods
    c) Avoid forecasts
    d) Use more chemicals
26. Which global body addresses food security and weather science?
    a) FAO
    b) WHO
    c) NASA
    d) WTO
27. WMO provides leadership in:
    a) Market prices
    b) Meteorological science and services
    c) Soil analysis
    d) Agroforestry
28. IPCC reports are crucial for:
    a) Banking sector
    b) Climate policy and action
    c) Fish export
    d) Fertilizer licensing
29. Agrometeorological research helps in
    a) Fuel development
    b) Improved climate resilience in crops
    c) Increased pesticide resistance
    d) Minimizing harvest size
30. Policy for food security must integrate:
    a) Weather and climate science
    b) Marketing agents
    c) Loan repayment strategies
    d) Branding efforts
31. Weather-smart agriculture includes
    a) Ignoring forecasts
    b) Rainfed monocropping
    c) Use of advisories, sensors, and resilient crops
    d) Dependence on old varieties

32. Soil moisture stress can be reduced by
    a) Heavy tillage
    b) Mulching and timely irrigation
    c) High fertilizer use
    d) Removing shade
33. Integrated farming reduces
    a) Weather risks to income
    b) Number of crops
    c) Government support
    d) Irrigation need
34. Diversification in cropping helps combat
    a) Soil erosion
    b) Single crop failure due to weather
    c) Water pollution
    d) Air pollution
35. Rainwater harvesting is a method to
    a) Reduce cloud formation
    b) Increase local humidity
    c) Store water during dry spells
    d) Avoid fertilization
36. A career in agricultural meteorology involves work with:
    a) Museums
    b) Meteorological services and agriculture departments
    c) Sports management
    d) Financial institutions
37. Agromet advisory preparation requires knowledge of:
    a) Carpentry
    b) Weather, crops, and local practices
    c) Banking rules
    d) Marketing skills
38. Agrometeorology helps in
    a) Predicting earthquakes
    b) Planning farm operations with weather insight
    c) Building dams
    d) Conducting veterinary surgery
39. Crop modeling under weather scenarios is used for
    a) TV programs
    b) Predicting yield and losses
    c) Air quality testing
    d) Advertising

40. Future food systems must rely on
    a) Traditional weather beliefs only
    b) Weather science and innovation
    c) Imports only
    d) Elimination of local crops
41. Resilient agriculture includes practices that
    a) Depend only on traditional knowledge
    b) Ignore forecast data
    c) Adapt to weather variability
    d) Promote water wastage
42. Early warning systems help in
    a) Price prediction
    b) Preparing for weather extremes
    c) Plant breeding
    d) Soil testing
43. Climate-smart agriculture integrates
    a) Traditional farming alone
    b) Livestock without feed planning
    c) Productivity with adaptation and mitigation
    d) Only irrigation techniques
44. ICT in agrometeorology helps in
    a) Weather manipulation
    b) Dissemination of weather-based advisories
    c) Fertilizer mixing
    d) Crop coloring
45. Seasonal forecast helps in
    a) Water harvesting
    b) Crop planning and input management
    c) Seed multiplication
    d) Irrigation design
46. Which factor does not directly influence weather?
    a) Latitude b) Altitude
    c) Fertilizer application d) Distance from sea

47. Weather parameters crucial during seed germination:
    a) Rainfall and wind speed
    b) Soil moisture and temperature
    c) Sunshine hours and cloud cover
    d) Fog and thunderstorm
48. Agrometeorological field units (AMFUs) provide
    a) Fertilizer subsidies
    b) Weather-based agromet advisories
    c) Seed distribution
    d) Land records
49. Weather's role in animal production is evident through:
    a) Improved meat color
    b) Influence on feed intake and productivity
    c) Better manure quality
    d) Ear size
50. Overall, weather serves agriculture by
    a) Reducing labor cost
    b) Guiding timely and efficient farm decisions
    c) Fixing prices
    d) Promoting imports

Answer Key

1	b	2	c	3	b	4	c	5	??	6	b	7	c
8	b	9	c	10	b	11	c	12	b	13	c	14	b
15	c	16	b	17	b	18	b	19	b	20	b	21	b
22	c	23	c	24	b	25	a	26	a	27	b	28	b
29	b	30	a	31	c	32	b	33	a	34	b	35	c
36	b	37	b	38	b	39	b	40	b	41	c	42	b
43	c	44	b	45	b	46	c	47	b	48	b	49	b
50	b												

Bibliography

Books

- Bishnoi, O.P. 2007. Principles of Agricultural Meteorology. Oxford Book Company. ISBN: 978- 8189473013. Pages 372.
- Mahi, G.A. & Kingra, P.K. 2018. Fundamentals of Agrometeorology and Climate Change. Kalyani Publishers. ISBN: 978-9327288995. Pages 383.
- Mavi, H.S. & Tupper, G.J. 2004. Agrometeorology: Principles and Applications of Climate Studies in Agriculture. CRC Press. ISBN: 978-1032768052. Pages 376.
- Mavi, HS. 1986. Introduction to Agrometeorology. Oxford & IBH Publishing Co Pvt. Ltd. ISBN: 978-8120409101. Pp 296.
- Murthy, V.R.K. 2015. Basic Principles of Agricultural Meteorology. BSP Books Private Limited. ISBN: 978-9352300747. Pages 288.
- Prasada Rao, G.S.L.H.V. 2008. Agricultural Meteorology. ISBN: 978-8120333383. PHI Learning Pvt Ltd. Pages 384.
- Ramana Rao, B.V. & Rao, V. Uma Maheswara Rao. 2024. Agricultural Meteorology. Jain Brothers. ISBN: 978-9390576975. Pages 154.
- Ramana Rao B.V., Surender Singh, V. Uma Maheswara Rao. 2025. Preparatory Microclimatology. Publisher: Professional Prints. New York USA. ISBN: 9781966695189 Pp 167.
- Ramana Rao B.V., Surender Singh, V. Uma Maheswara Rao. 2025. Conceptual Meteorology. Publisher: New India Publishing Agency, New Delhi. ISBN 978-9358874501 Pp 152.
- Ramana Rao B.V., Surender Singh, V. Uma Maheswara Rao. 2025. Primer on Agrohydrometeorology. Publisher: Professional Prints. New York USA. ISBN: 978-1966695301 Pp 144.
- Ramana Rao B.V., Surender Singh, V. Uma Maheswara Rao, Anu Bulusu. 2025. Farming Weather: Harnessing Citizen Science for Smart Agriculture. Publisher: Professional Prints. New York USA. ISBN: 978-1966695561. Pp 154.
- Ramana Rao B.V., Surender Singh, V. Uma Maheswara Rao. 2025. Forecasting the Future: Effective Climate Guidance on Monsoons for Farm Prosperity. Publisher: Professional Prints. New York USA. ISBN: 978-1966695912. Pp 246.
- Ramana Rao B.V., Surender Singh, V. Uma Maheswara Rao. 2025. Preparatory Microclimatology. Publisher: Professional Prints. New York USA. ISBN: 978-1966695189 Pp 167.
- Ramana Rao B.V., Surender Singh, V. Uma Maheswara Rao. 2025. Conceptual Meteorology. Publisher: New India Publishing Agency, New Delhi. ISBN 978-9358874501 Pp 152.

- Ramana Rao B.V., Surender Singh, V. Uma Maheswara Rao. 2025. Primer on Agrohydrometeorology. Publisher: Professional Prints. New York USA. ISBN: 978-1966695301 Pp 144.
- Ramana Rao B.V., Surender Singh, V. Uma Maheswara Rao, Anu Bulusu. 2025. Farming Weather: Harnessing Citizen Science for Smart Agriculture. Publisher: Professional Prints. New York USA. ISBN: 978-1966695561. Pp 154.
- Ramana Rao B.V., Surender Singh, V. Uma Maheswara Rao. 2025. Forecasting the Future: Effective Climate Guidance on Monsoons for Farm Prosperity. Publisher: Professional Prints. New York USA. ISBN: 978-1966695912. Pp 246.
- Seemann, J.; Chirkov, Y.I.; Lomas, J. & Primault, B. 1979. Agrometeorology. Springer Berlin, Heidelberg. ISBN: 978-3642672903. Pages 326.
- Varshneya, M.C. & Pillai, B.P. 2004. Textbook of Agricultural Meteorology. ICAR Publication. ISBN: 978-8171640195. Pages 242.

Journals

- Agricultural and Forest Meteorology. https://www.sciencedirect.com/journal/agricultural-and-forest-meteorology
- Chinese Journal of Agrometeorology. https://zgnyqx.ieda.org.cn/EN/1000-6362/home.shtml
- Italian Journal of Agrometeorology. https://riviste.fupress.net/index.php/IJAm/issue/view/15
- Journal of Agril Meteorology. :https://journals.indexcopernicus.com/search/details? id =46231
- Journal of Agrometeorology. https://journal.agrimetassociation.org/index.php/jam/issue/view/73
- MAUSAM. https://mausamjournal.imd.gov.in/index.php/MAUSAM

Web Resources

- https://mausam.imd.gov.in/
- https://imdagrimet.gov.in/
- https://www.tropmet.res.in/
- https://nwp.ncmrwf.gov.in/
- https://www.skymetweather.com/
- https://icar.org.in/
- https://www.icar-crida.res.in/
- https://niasm.icar.gov.in/
- https://www.sac.gov.in/Vyom/
- https://www.isro.gov.in/
- https://www.icrisat.org/
- https://www.ecmwf.int/
- https://wmo.int/
- https://www.fao.org/climate-smart-agriculture/en/
- https://www.agrimetassociation.org/
- https://safoam.org.in/
- https://www.insam.org.in/

Subject Index

S

T